Transition-Metal Organometallic Chemistry
An Introduction

TRANSITION-METAL
ORGANOMETALLIC CHEMISTRY
An Introduction

R. Bruce King

Department of Chemistry
University of Georgia
Athens, Georgia

ACADEMIC PRESS   New York and London 1969

ACADEMIC PRESS, INC.
111 Fifth Avenue, New York, New York 10003

United Kingdom Edition published by
ACADEMIC PRESS, INC. (LONDON) LTD.
Berkeley Square House, London W1X 6BA

LIBRARY OF CONGRESS CATALOG CARD NUMBER: 72-84240

PRINTED IN THE UNITED STATES OF AMERICA

PREFACE

Since the discovery of ferrocene in 1951, transition-metal organometallic chemistry has developed into a major discipline. Many compounds with unusual structures, reactions, and physical properties have been discovered. Such compounds are of interest both in various areas of basic chemistry and in certain applications, notably catalysis and metal deposition.

The growing body of information in this discipline, as well as its increasing relevance to most areas of chemistry, makes some knowledge of this field necessary for many practicing chemists in other areas. This book presents the basic facts and principles of transition-metal organometallic chemistry in a manner suitable to introduce this field to graduate students, advanced undergraduates, and practicing research workers. Although the entire field is not covered exhaustively, extensive references are given for the reader wishing more detailed information on various subjects discussed in the book.

The first chapter is the longest one since it summarizes the most important principles in the field. The later chapters are more descriptive in nature, summarizing specific organometallic chemistry of the various transition metals according to their positions in the Periodic Table. This type of organization not only permits emphasis of general principles such as might be desirable for a course, but also makes this book suitable for an introduction to the organometallic chemistry of a specific metal which may be of particular interest to the reader.

The material for this book is based on a short course sponsored by the American Chemical Society which I presented in Atlanta, Georgia, Bethlehem, Pennsylvania, and Boston, Massachusetts at various times during 1967 and 1968. It should be useful as a text for a portion of a graduate level course in organometallic chemistry. It is directed to a reader at the level of a beginning graduate student with a good background in all areas of undergraduate chemistry but no prior specific background in transition-metal organometallic chemistry. However, many active research workers in one area of this field may also find this work a useful review for other areas of the discipline.

I would like to acknowledge helpful discussions with Professor E. C. Ashby of the Georgia Institute of Technology with whom I gave the short course mentioned above. I would also like to acknowledge the cooperation of my wife and family during the various phases of preparing this book.

Athens, Georgia   R. Bruce King
July, 1969

v

Contents

Transition-Metal
Organometallic Chemistry
An Introduction

CHAPTER I

General Principles of Transition-Metal Organometallic Chemistry

Introduction

Transition metals are defined as elements which possess incompletely filled d orbitals in the ground state. They thus include the metals shown in the portion of the periodic table depicted in Table I-1. All of these metals form organometallic derivatives. However, with one exception, presently known organometallic derivatives of the lanthanides and actinides are limited to highly ionic derivatives of the cyclopentadienide and substituted cyclopentadienide anions.

The availability of low-lying, filled metal d orbitals for participation in the bonding makes transition-metal organometallic derivatives very different both in type and properties from nontransition-metal organometallic derivatives. The nontransition-metal organometallic compounds normally have "conventional" organic groups, such as methyl, ethyl, and phenyl, bonded to the metal atom by σ-bonds similar in type to the carbon–carbon and carbon–hydrogen bonds commonly encountered in organic chemistry. The only significant deviations from this pattern in nontransition-metal organometallic chemistry occur in the highly ionic bonds found in alkyls and aryls of the alkali metals and other very electropositive elements, and in the electron-deficient alkyl bridge bonds found in aluminum alkyls and related compounds. By contrast, transition-metal organometallic compounds contain a much greater variety of organic groups, often with much more complicated (and sometimes ill-understood) bonding schemes frequently involving the d electrons of the transition metals. Many of the most common organic groups in transition-metal organometallic chemistry, including the ubiquitous carbonyl and π-cyclopentadienyl groups, are not encountered at all in nontransition-metal organometallic chemistry. The first chapter in this book on transition-metal organometallic chemistry will discuss, in a qualitative and somewhat over-simplified manner, the bonding to transition metals of some of the most

1

TABLE I-1

THE TRANSITION METALS

Group	III	IV	V	VI	VII	VIII	VIII	VIII	IB
Electrons needed to attain rare-gas configuration	15	14	13	12	11	10	9	8	7
3d Transition series	Sc Scandium Z = 21	Ti Titanium Z = 22	V Vanadium Z = 23	Cr Chromium Z = 24	Mn Manganese Z = 25	Fe Iron Z = 26	Co Cobalt Z = 27	Ni Nickel Z = 28	Cu Copper Z = 29
4d Transition series	Y Yttrium Z = 39	Zr Zirconium Z = 40	Nb Niobium Z = 41	Mo Molybdenum Z = 42	Tc Technetium Z = 43	Ru Ruthenium Z = 44	Rh Rhodium Z = 45	Pd Palladium Z = 46	Ag Silver Z = 47
5d Transition series	Lanthanides La Ce Pr Nd Pm Sm Eu Gd Tb Dy Ho Er Tm Yb Lu	Hf Hafnium Z = 72	Ta Tantalum Z = 73	W Tungsten Z = 74	Re Rhenium Z = 75	Os Osmium Z = 76	Ir Iridium Z = 77	Pt Platinum Z = 78	Au Gold Z = 79

important organic groups forming metal–carbon bonds. However, in order to give the reader an approximate indication of the development of transition-metal organometallic chemistry, a brief historical survey of some of the key developments in this area will first be presented.

Historical Development of Transition-Metal Organometallic Chemistry

The development of transition-metal organometallic chemistry dates back to 1827 when Zeise [1] reported the first transition-metal organometallic compound, the ethylene–platinum complex $K[PtCl_3C_2H_4]$. Subsequent developments in this area of organometallic chemistry arose not in orderly steps from this original discovery but, instead, from several other initially unrelated discoveries. These include the discovery of nickel tetracarbonyl in 1890 by Mond, Langer, and Quincke [2] which led to the area of metal carbonyl chemistry developed further largely by Hieber and his co-workers from about 1920 to the present. Another initially unrelated discovery was that of the polyphenylchromium compounds by Hein in 1919 [3]. These compounds represented one of the great enigmas in inorganic chemistry for several decades until Fischer and Hafner [4] discovered dibenzenechromium in 1955. Subsequently the bonding in dibenzenechromium was shown to be similar to that in the polyphenylchromium compounds of Hein. Meanwhile, another key event in transition-metal organometallic chemistry had occurred: the discovery of ferrocene (biscyclopentadienyliron) in 1951 by two different groups of workers (Kealy and Pauson [5] in the United States and Miller, Tebboth, and Tremaine [6] in Great Britain). The discovery of ferrocene led to the development of metal cyclopentadienyl chemistry largely in two different laboratories: that of Wilkinson at Harvard University (United States) and that of E. O. Fischer in Munich, Germany. Since the middle 1950's the synthesis and study of unusual transition-metal organometallic derivatives has continued at a fast and furious rate in many laboratories throughout all developed nations of the world. Many of these studies have been of an intensive nature, concentrating on a specific transition-metal organometallic system, often for achieving a specific scientific or practical objective. However, a few laboratories, including particularly that of E. O. Fischer, have been actively pursuing extensive research programs in transition-metal organometallic chemistry involving almost all available transition metals and organic groups capable of forming metal–carbon bonds of one type or another.

Transition-metal organometallic compounds have been found to have certain applications of practical or potentially practical interest. Some types have catalytic applications, such as certain organotitanium systems acting as components of Ziegler–Natta polymerization catalysts, certain organo-rhodium systems serving as catalysts for the preparation of unusual olefins (e.g., 1,4-hexadiene), and organopalladium compounds acting as inter-mediates in the palladium-catalyzed oxidation of olefins. Two metal carbonyls manufactured in tonnage quantities are iron pentacarbonyl, from which special iron powders are prepared by thermal decomposition, and methyl-cyclopentadienylmanganese tricarbonyl, for use as an additive to improve the combustion of certain liquid fuels. Further applications of metal carbonyls as vehicles for the transport and deposition of metals are exemplified by the role of nickel tetracarbonyl in the Mond process for nickel refining and by the use of the thermal decomposition of molybdenum hexacarbonyl as a means of depositing metallic molybdenum.

At the present time, many transition-metal organometallic compounds, including most of the metal carbonyls, metal cyclopentadienyls, and cyclo-pentadienyl metal carbonyls are commercially available at prices ranging from one to two dollars per pound for some products produced in tonnage quantities, such as iron pentacarbonyl and methylcyclopentadienylmanganese tricarbonyl, to hundreds of dollars per gram for difficultly preparable compounds and derivatives of rare metals, such as ditechnetium decacarbonyl. This makes much significant research in transition-metal organometallic chemistry pos-sible with equipment hardly more complicated and unusual than that cus-tomarily used for much research in more traditional areas of organic chemistry. The main difference is that many laboratory operations which would be done in air in conventional organic chemistry must often be done under an inert atmosphere (almost always nitrogen) in transition-metal organometallic chemistry.

Some Aspects of the Bonding in Coordination Compounds

Transition-metal organometallic compounds may be regarded as special types of coordination compounds. In understanding the various types of transition-metal organometallic compounds, it is first useful to consider some of the aspects of bonding in coordination compounds. The ligands commonly found in transition-metal organometallic compounds, such as carbonyls, cyclopentadienyls, and olefins, are strong-field ligands (such as cyanide in potassium ferrocyanide, etc.). Such strong-field ligands can cause electrons in the free transition-metal ion to pair, generally resulting in complexes with a minimum of unpaired electrons. A recently reported [7] formal scheme for organizing coordination compounds is particularly appropriate for deriva-tives with strong-field ligands, such as transition-metal organometallic com-pounds. The following discussion will use the formalism of this recently reported scheme where appropriate.

Two concepts crucial to the understanding of coordination compounds are electronic configuration and coordination number. Both of these concepts, although inherently simple, may acquire certain complexities when applied to certain types of transition-metal organometallic compounds. In complexes with strong-field ligands, such as transition-metal organometallic derivatives, attainment of a favorable electronic configuration, generally the eighteen-(outer-) electron configuration of the next rare gas, takes precedence over attainment of a favorable coordination number, such as 4 or 6. This contrasts with the case of transition-metal complexes of weak-field ligands, where attainment of a favorable coordination number (generally 4 or 6) takes precedence over attainment of a favorable electronic configuration.

Consideration of the electronic configuration of a transition-metal organometallic derivative is clearest if both the metal atom and the ligands are regarded as neutral species [7]. Ligands encountered in transition-metal organometallic chemistry can donate from zero to eight electrons to the metal atom. Examples of ligands donating various numbers of electrons to the central metal atom are listed in Table I-2.

TABLE I-2

EXAMPLES OF LIGANDS DONATING VARIOUS NUMBERS OF ELECTRONS

Number of electrons donated to metal	Examples of ligands donating this number of electrons
0	BH_3, O (oxide)
1	Halides, Alkyl and Aryl Groups, R_3Sn, H, SCN, NO_2, NO_3
2	C_2H_4, CO, R_3N, R_3P, R_3As, RCN, RNC, R_2O, R_2S
3	π-C_3H_5 (allyl), NO, RN_2
4	π-C_4H_4 (cyclobutadiene), π-C_4H_6 (butadiene)
5	π-C_5H_5 (cyclopentadienyl), π-C_9H_7 (indenyl)
6	π-C_6H_6 (benzene), π-C_7H_8 (cycloheptatriene)
7	π-C_7H_7 (cycloheptatrienyl)
8	π-C_8H_8 (cyclooctatetraene)—very rare

In calculating the electronic configuration of a metal atom in a complex, the total number of electrons donated by all of the (neutral) ligands is added to the number of electrons of the neutral metal atom. In most transition-metal organometallic derivatives, the metal atom will be found to attain the electronic configuration of the next rare (inert) gas (krypton, xenon, or radon). Examples of such complexes with the electronic configuration of the next rare gas and illustrations of the method of calculation of the electronic configuration are given in Table I-3. A few examples of complexes with electronic configurations other than that of the next rare gas are given in Table I-4. In Tables I-3 and I-4, only the electrons in the outer, incomplete shells of the metal atoms are considered, which makes eighteen (outer) electrons the configuration of the next inert gas, whether krypton, xenon, or radon.

The second concept of importance is that of coordination number. In complexes containing exclusively monodentate ligands, the coordination number is the number of ligands bonded to the metal atom. In the cases of

TABLE I-3

EXAMPLES OF ELECTRONIC CONFIGURATION CALCULATIONS

(A)	$Ni(CO)_4$	
	Neutral nickel atom	10
	Four carbonyl groups	$4 \times 2 =$ 8
	Electronic configuration	18
(B)	$Mn_2(CO)_{10}$	
	Neutral manganese atom	7
	Five carbonyl groups (per manganese atom)	$5 \times 2 = 10$
	Manganese–manganese covalent bond	1
	Electronic configuration	18
(C)	$C_5H_5(CO)_4$	
	Neutral vanadium atom	5
	Four carbonyl groups	$4 \times 2 =$ 8
	π-Cyclopentadienyl ring	5
	Electronic configuration	18
(D)	$C_7H_7W(CO)_2I$	
	Neutral tungsten atom	6
	Two carbonyl groups	$2 \times 2 =$ 4
	Iodine atom	1
	π-Cycloheptatrienyl ring	7
	Electronic configuration	18
(E)	π-$C_3H_5Fe(CO)_2NO$	
	Neutral iron atom	8
	Two carbonyl groups	$2 \times 2 =$ 4
	Nitrosyl group	3
	π-Allyl ligand	3
	Electronic configuration	18

TABLE I-4

FURTHER EXAMPLES OF ELECTRONIC CONFIGURATION CALCULATIONS

(F)	$V(CO)_6$	
	Neutral vanadium atom	5
	Six carbonyl groups	$6 \times 2 = 12$
	Electronic configuration	17
(G)	$(C_5H_5)_2Ni$	
	Neutral nickel atom	10
	Two π-cyclopentadienyl ligands	$2 \times 5 = 10$
	Electronic configuration	20
(H)	$(R_3P)_2IrCOCl$	
	Neutral iridium atom	9
	One carbonyl group	2
	Two R_3P ligands	$2 \times 2 = 4$
	Chlorine atom	1
	Electronic configuration	16

polydentate ligands, it is necessary to count the "arms" of the ligands rather than the number of individual ligands. Delocalized ligands including cyclic C_nH_n systems, such as π-cyclopentadienyl, can generally be considered as bonded to the metal atom by donation of two or three electron pairs. Such delocalized ligands may be considered as formally bidentate or tridentate

TABLE I-5

EXAMPLES OF TRANSITION METAL ORGANOMETALLIC DERIVATIVES WITH DIFFERENT COORDINATION NUMBERS

Coordination number	Examples
3	$C_{12}H_{18}Ni$ (three nickel–olefin bonds)
4	$Ni(CO)_4$, C_5H_5NiNO, $Co(CO)_3)NO$, $Co(CO)_4^-$, $Fe(CO)_3NO^-$
5	$Fe(CO)_5$, $R_3SnCo(CO)_4$, $C_5H_5Co(CO)_2$, $C_5H_5Rh(C_2H_4)_2$
6	$Cr(CO)_6$, $C_5H_5Mn(CO)_3$, $RMn(CO)_5$, $(C_5H_5)_2Fe$, $(C_6H_6)_2Cr$, $C_5H_5Fe(CO)_2R$, $(C_6H_5)_3PV(CO)_4NO$, $C_5H_5Mo(CO)_2NO$
7	$C_5H_5V(CO)_4$, $RHgTa(CO)_6$, $[Mo(CO)_4X_3]^-$, $[C_5H_5Mo(CO)C_6H_6]^+$
8	$C_5H_5Mo(CO)_2I_3$, $Mo(CN)_8^{4-}$
9	ReH_9^-, $[C_5H_5FeCO]_4$

ligands, respectively. Table I-5 gives examples of transition-metal organo-metallic derivatives with different coordination numbers (from 3 to 9) of the central metal atom.

When considering transition-metal organometallic compounds, it is convenient to classify the carbon ligands according to the number of their carbon atoms bonded to one (or sometimes two or more) metal atoms. Examples of this classification of carbon ligands are given in Table I-6. Ligands bonding

TABLE I-6

CLASSIFICATION OF CARBON LIGANDS BASED ON NUMBER OF CARBON ATOMS BONDED TO METAL ATOM(S)

Number of carbon atoms	Number of metal atoms	Ligands
1	1	CO, CH_3, CF_3, C_6H_5, CNR
2	1	Ethylene
3	1	π-Allyl
4	1	π-Butadiene, π-cyclobutadiene, trimethylenemethane
5	1	π-Cyclopentadienyl, π-indenyl, π-cyclohexadienyl
6	1	π-Benzene, π-cycloheptatriene
7	1	π-Cycloheptatrienyl
8	1	π-C_8H_8 in $(C_8H_8)_3Ti_2$
1	2	Bridging CO [e.g., $Co_2(CO)_8$]
2	2	Bridging acetylene [e.g., $R_2C_2Co_2(CO)_6$]
3	2	
4	2	Butatriene [e.g., $C_4H_4Fe_2(CO)_6$]
5	2	
6	2	Bisallylene [e.g., $C_6H_8Fe_2(CO)_6$]
1	3	Three-way bridging CO [e.g., $(C_5H_5)_3Ni_3(CO)_2$]

to a single metal atom by means of one carbon atom include carbon monoxide, various isocyanides, and various alkyl, fluoroalkyl, aryl, fluoroaryl, and acyl groups. Ethylene and substituted olefins may bond to a single transition-metal atom by means of two carbon atoms. The π-allyl group is a V-shaped array of three carbon atoms which can bond to a single metal atom. The related triangular array of three carbon atoms (the π-cyclopropenyl ligand) is also, in principle, capable of bonding to a single metal atom. Ligands where four carbon atoms can bond to a single metal atom include cases where the four carbon atoms form a rectangle (cyclobutadiene), chain (butadiene), or Y (trimethylenemethane). Ligands in which five to eight carbon atoms can bond to a single metal atom include, particularly, the regular C_nH_n polygons. Such important ligands as π-cyclopentadienyl and benzene fall into this category.

Carbon ligands are also encountered in transition-metal organometallic chemistry where one or more carbon atoms bond simultaneously to two or more metal atoms. Thus the single carbon atom in a carbonyl group can bond simultaneously to two or three metal atoms forming metal derivatives with bridging carbonyl groups. Similarly, in compounds of the type $R_2C_2Co_2(CO)_6$, the two carbon atoms of an acetylene derivative can bond simultaneously to two metal atoms. Extreme examples of a single carbon atom bonding to many metal atoms occur in the carbonyl carbides $Fe_5(CO)_{15}C$ and $Ru_6(CO)_{17}C$, where the single carbon atom bonds simultaneously to five iron atoms [8] and six ruthenium atoms [9], respectively. In both of these cases, the single carbon atom is located in the center of a polyhedron of the metal atoms.

A characteristic feature of transition-metal atoms with coordination numbers less than nine is the availability of electron pairs in d orbitals not involved in the bonds formed by electron donation from the ligands to the metal atom. These filled metal d orbitals can overlap with the empty antibonding orbitals in many of the ligands commonly encountered in transition-metal organometallic chemistry. This additional bond strengthens the metal–ligand linkage. However, since this additional bonding places additional electrons in ligand antibonding orbitals, it weakens certain bonds in the ligand. This additional bonding gives transition-metal organometallic chemistry many of its distinctive features, and is known colloquially as "back bonding" or, more precisely, as "retrodative π-bonding."

Nontransition metals do not have filled d orbitals of appropriate energies and, therefore, cannot participate in retrodative π-bonding. The phenomenon of retrodative π-bonding accounts for the unique ability of transition metals with available filled d orbitals to form stable complexes with the extremely weak Lewis base carbon monoxide. The retrodative π-bonding of this type is important in metal carbonyls, olefin complexes, and derivatives of certain tricovalent phosphorus compounds, notably PF_3. Analogous retrodative δ-bonding occurs in π-cyclopentadienyl derivatives and similar compounds; this will be discussed briefly when the π-C_5H_5 ligand is treated.

Metal Carbonyls

BONDING

One of the most commonly encountered ligands in transition-metal organometallic chemistry is carbon monoxide which forms complexes with these metals known as the metal carbonyls. Figure I-1 depicts the salient features of the bond between transition metals and carbon monoxide. The carbon monoxide has lone electron pairs on both the carbon and oxygen atoms. The sp orbital of the carbon atom containing its lone electron pair can overlap with a metal hybrid orbital to form a forward σ-bond, where this lone electron pair of the ligand is donated to the metal atom. This bond is supplemented by a second bond, a reverse π-bond, where a filled metal d orbital containing an electron pair can overlap with an empty π^* antibonding orbital of the carbon monoxide ligand. The extent of this retrodative π-bonding is quite variable in metal carbonyl derivatives, depending upon the electron density on the metal atom which is affected by various factors, such as the other ligands attached to the metal atom and the charge on the metal carbonyl species. In the case of metal carbonyls, the forward σ-bond between the transition metal and carbon monoxide is sufficiently weak that significant amounts of retrodative π-bonding

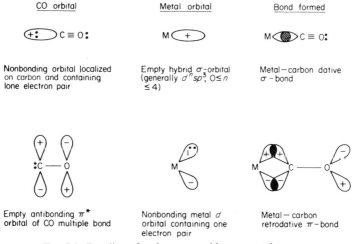

FIG. I-1. Bonding of carbon monoxide to a metal atom.

are required to insure reasonable stability of the compound. The following additional features of the reverse (π) bond in metal carbonyls are also of some importance. (a) Carbon monoxide has *two* orthogonal antibonding π^* orbitals which can form *two* orthogonal reverse π-bonds with the metal atom, provided that suitable metal orbitals are available. (b) The retrodative π-bonding in metal carbonyls places additional electron density into the antibonding π^* orbitals of the carbon monoxide. This lowers the carbon–oxygen bond order and, therefore, the carbon–oxygen stretching force constant, an effect that can be measured from infrared frequencies. Thus, infrared spectroscopic studies of carbon–oxygen stretching frequencies in metal carbonyls provide a useful and frequently used means for deducing the amount of retrodative π-bonding.

TYPES OF COMPOUNDS FORMED

Table I-7 lists the known neutral transition-metal derivatives containing only carbonyl groups. Almost all of these compounds have the favored

TABLE I-7

METAL CARBONYLS

Group	III	IV	V	VI	VII	VIII	VIII	VIII	VIII	IB
Electrons needed to attain rare-gas configuration	15	14	13	12	11	10	9	8		7
3d Transition series			$V(CO)_6$ blue cryst.	$Cr(CO)_6$ white cryst.	$Mn_2(CO)_{10}$ yellow cryst.	$Fe(CO)_5$ yellow liq. $Fe_2(CO)_9$ yellow-orange cryst. $Fe_3(CO)_{12}$ black cryst.	$Co_2(CO)_8$ orange cryst. $Co_4(CO)_{12}$ black cryst. $Co_6(CO)_{16}$ black cryst.	$Ni(CO)_4$ colorless liq.		
4d Transition series				$Mo(CO)_6$ white cryst.	$Tc_2(CO)_{10}$ white cryst.	$Ru(CO)_5$ colorless liq. $Ru_3(CO)_{12}$ yellow-orange cryst.	$Rh_4(CO)_{12}$ red-orange cryst. $Rh_6(CO)_{16}$ black cryst.			
5d Transition series				$W(CO)_6$ white cryst.	$Re_2(CO)_{10}$ white cryst.	$Os(CO)_5$ colorless liq. $Os_3(CO)_{12}$ yellow cryst.	$Ir_4(CO)_{12}$ yellow cryst.	$[Pt(CO)_2]_n$ brown solid		

eighteen-electron rare-gas electronic configuration. A notable exception is hexacarbonylvanadium, $V(CO)_6$, which has only a seventeen-electron configuration and, therefore, is paramagnetic to the extent expected for one unpaired electron. Hexacarbonylvanadium is much less stable and much more reactive than the isostructural hexacarbonylchromium, $Cr(CO)_6$, which does have the favored rare-gas configuration.

PREPARATIVE METHODS

In a few cases, metal carbonyl derivatives can be made by direct reaction of gaseous carbon monoxide with the free metal, generally in a finely divided pyrophoric form. Thus nickel tetracarbonyl, $Ni(CO)_4$, may be made by direct reaction of finely divided nickel metal with carbon monoxide at atmospheric pressure [10]. The ability of nickel, but not cobalt or other metals, to undergo this reaction with carbon monoxide at atmospheric pressure is the basis for the Mond process for the refining of nickel.

In the cases of metals other than nickel, direct reaction with carbon monoxide does not occur at atmospheric pressure. However, in the cases of iron, cobalt, molybdenum, and ruthenium, combination with carbon monoxide can be effected at elevated temperatures and pressures to give the carbonyls $Fe(CO)_5$, $Co_2(CO)_8$, $Mo(CO)_6$, and $Ru(CO)_5$, respectively. Good yields of iron and cobalt carbonyls can be achieved by this direct combination method, but in the cases of molybdenum and ruthenium, the yields of the carbonyls from the direct combination of the metal with carbon monoxide are very poor. Other less direct methods described below and in the later chapters dealing with these metals are preferable.

A much more useful method for the preparation of most metal carbonyls is the reductive carbonylation reaction. This reaction is based on the reaction of a metal compound with carbon monoxide in the presence of a reducing agent at elevated temperatures and pressures. The selection of a reducing agent is often a rather delicate matter, since the reducing agent must be strong enough to effect the reaction without converting the metal derivative directly to the free metal (without bonding of the carbon monoxide). A wide variety of reducing agents and systems have been utilized for this reductive carbonylation reaction. Some of the more useful reducing agents are excess carbon monoxide, hydrogen, sodium, magnesium, aluminum, zinc, copper, alkylmagnesium halides (Grignard reagents), aluminum alkyls, lithium aluminum hydride, and sodium benzophenone ketyl. The following examples illustrate the applications of some of these reagents.

a. *Excess carbon monoxide [11]*:

$$Re_2O_7 + 17\ CO \xrightarrow[350\ atm]{250°} Re_2(CO)_{10} + 7\ CO_2$$

$$OsO_4 + 9\ CO \longrightarrow Os(CO)_5 + 4\ CO_2$$

b. *Hydrogen [12]:*

$$2 \text{ CoCO}_3 + 8 \text{ CO} + 2 \text{ H}_2 \longrightarrow \text{Co}_2(\text{CO})_8 + 2 \text{ CO}_2 + 2 \text{ H}_2\text{O}$$

An intermediate in this reaction is the hydrocarbonyl HCo(CO)_4.

c. *Sodium [13]:*

$$\text{VCl}_3 + 4 \text{ Na} + 2 \text{ diglyme} + 6 \text{ CO} \xrightarrow[\text{150 atm}]{100°} [\text{Na(diglyme)}_2] [\text{V(CO)}_6] + 3 \text{ NaCl}$$

Analogous synthetic techniques can be used to prepare the niobium and tantalum analogues to this vanadium compound as well as the neutral carbonyls M(CO)_6 (M = Cr, Mo, and W) and $\text{Re}_2(\text{CO})_{10}$.

d. *Magnesium [14]:* reaction of chromium (III) acetylacetonate in pyridine solution with magnesium in the presence of carbon monoxide under pressure and catalytic amounts of iodine gives a good yield of Cr(CO)_6.

e. *Aluminum [15]:*

$$\text{CrCl}_3 + \text{Al} + 6 \text{ CO} \xrightarrow[\text{AlCl}_3]{\text{benzene}} \text{Cr(CO)}_6 + \text{AlCl}_3$$

A dibenzenechromium derivative is an intermediate in this reaction.

f. *Zinc [16]:*

$$\text{WCl}_6 + 3 \text{ Zn} + 6 \text{ CO} \longrightarrow \text{W(CO)}_6 + 3 \text{ ZnCl}_2$$

g. *Copper [17]:*

$$6 \text{ RhCl}_3 + 18 \text{ Cu} + 16 \text{ CO} \longrightarrow \text{Rh}_6(\text{CO})_{16} + 18 \text{ CuCl}$$

$$\text{K}_2\text{ReX}_6 + 3 \text{ Cu} + 5 \text{ CO} \longrightarrow \text{Re(CO)}_5\text{X} + 2 \text{ KCl} + 3 \text{ CuCl}$$

where X = Cl, Br, or I.

h. *Alkylmagnesium halides [18]:* the metal hexacarbonyls M(CO)_6 (M = Cr, Mo, and W) were originally prepared by treatment of the metal halide with carbon monoxide in the presence of a diethyl ether solution of an alkylmagnesium halide.

i. *Aluminum alkyls [19]:* one of the more useful methods of preparing $\text{Mn}_2(\text{CO})_{10}$ consists of the reaction of manganese (II) acetate with excess triisobutylaluminum in diisopropyl ether in the presence of carbon monoxide under pressure.

j. *Lithium aluminum hydride [20]:* lithium aluminum hydride (LiAlH_4) in diethyl ether solution has been used to convert the halides of chromium, molybdenum, and tungsten into their hexacarbonyls M(CO)_6 (M = Cr, Mo, and W) in the presence of carbon monoxide under pressure.

k. *Sodium benzophenone ketyl [21]:* reaction of benzophenone with metallic sodium in tetrahydrofuran solution gives the blue ketyl $\text{Na}[(\text{C}_6\text{H}_5)_2\text{CO}]$. Treatment of this ketyl with anhydrous manganese (II) chloride followed by reaction with carbon monoxide at $\sim200°/200$ atm provided the first method for preparing appreciable quantities of $Mn_2(CO)_{10}$.

Further discussion of these specific reactions will be deferred until the sections on the specific metals involved.

In rare cases, iron pentacarbonyl rather than carbon monoxide can be used as the source of carbon monoxide in the preparation of metal carbonyls. Thus, tungsten hexacarbonyl can be prepared in reasonable yield by treatment of tungsten hexachloride with iron pentacarbonyl in an autoclave containing hydrogen but *not* carbon monoxide under pressure [22].

Metal Cyclopentadienyls

BONDING

Another frequently encountered ligand in transition-metal organometallic chemistry is the cyclopentadienide anion which forms complexes known as the metal cyclopentadienyls with these metals. In most of these compounds all five carbon atoms of the cyclopentadienyl ring are bonded to the metal atom. Compounds of this type are known as the π-cyclopentadienyls. A very stable and familiar compound of this type is biscyclopentadienyliron, $(C_5H_5)_2Fe$, which has acquired the more convenient trivial name ferrocene.

Figure I-2 depicts the features of the bond between the transition metal and the cyclopentadienyl ring in the π-cyclopentadienyl derivatives. In order to understand this type of bonding, it is first necessary to consider the molecular orbitals of the symmetrical, planar, pentagonal C_5H_5 ring. These can be divided into the following three types:

a. *A bonding orbital*—this orbital has no nodes, and, in the $C_5H_5^-$ anion, contains an electron pair.

b. E_1 *bonding orbitals*—these orbitals represent a degenerate pair. Each orbital has a single node. The two nodes in the two orbitals of this degenerate pair are perpendicular, i.e., they form 90° angles with one another. In the $C_5H_5^-$ anion, each of the two degenerate E_1 bonding orbitals contains an electron pair.

c. E_2 *antibonding orbitals*—these orbitals, like the E_1 bonding orbitals, represent a degenerate pair. Each E_2 orbital has two perpendicular nodes. The two pairs of perpendicular nodes in the two orbitals of this degenerate pair form 45° angles with one another. In the $C_5H_5^-$ anion, neither of the two degenerate E_2 antibonding orbitals contains an electron pair.

The $C_5H_5^-$ anion in π-cyclopentadienylmetal derivatives may be regarded as a tridentate ligand, i.e., a ligand that donates three electron pairs to the metal atom. The filled A bonding orbital of the $C_5H_5^-$ ring can donate its electron pair to the metal atom by forming a σ-bond with a metal hybrid

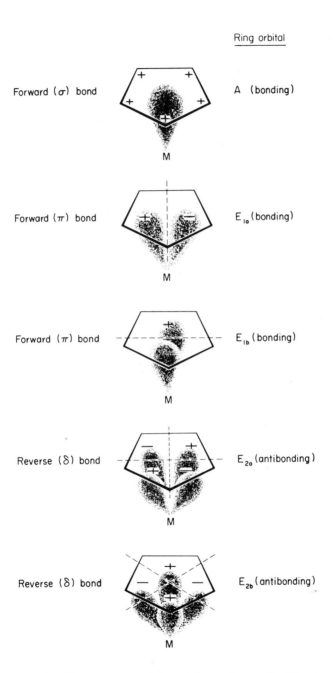

FIG. I-2. The components of the metal–π-cyclopentadienyl bond.

orbital. The pair of orthogonal filled E_1 bonding orbitals of the $C_5H_5^-$ ring can donate their two electron pairs to the metal atom by forming a pair of orthogonal π-bonds with appropriate metal orbitals. The $C_5H_5^-$ ring thus donates its three electron pairs to the metal atom by means of a metal–ring triple bond composed of one σ-bond and two orthogonal π-bonds similar to the carbon–carbon triple bond in acetylene and its derivatives. Furthermore, the metal–ring triple bond in π-cyclopentadienylmetal derivatives with only one σ-bond and two orthogonal π-bonds uses different metal orbitals than the three σ-bonds between metal atoms and either of three monodentate ligands such as carbonyl groups or a nondelocalized tridentate ligand such as diethyl-enetriamine. The differences in the metal orbitals involved in the forward bonding mean that in the hexacovalent derivative $Cr(CO)_6$ the forward bonding uses the s, p_x, p_y, p_z, d_{z2}, and d_{x2-y2} metal orbitals, whereas in the similarly formal hexacovalent derivatives $C_5H_5Mn(CO)_3$ and $(C_5H_5)_2Fe$ the forward bonding uses the s, p_x, p_y, p_z d_{xz}, and d_{yz} metal orbitals.

In the discussion on the bonding in metal carbonyls earlier in this chapter, the occurrence of retrodative bonding was mentioned. Such retrodative bonding also occurs in π-cyclopentadienylmetal derivatives, but to a lesser extent than in the metal carbonyls. The retrodative bonding in π-cyclopenta-dienylmetal derivatives involves partial back donation of the electron pair in a binodal metal orbital not involved in the forward bonding (generally the d_{x2-y2} and d_{xy} orbitals) to the empty binodal ligand E_2 orbitals. Since these reverse metal–ligand bonds in π-cyclopentadienylmetal compounds have two perpendicular nodal planes, this type of back bonding in π-cyclopentadienyl-metal derivatives may be regarded as retrodative δ-bonding. This contrasts with the reverse metal–ligand bonds in metal carbonyls, which have but one nodal plane and, therefore, represent retrodative π-bonding.

TYPES OF COMPOUNDS FORMED

Table I-8 lists the known transition-metal derivatives containing just π-cyclopentadienyl ligands. Similarly, Table I-9 lists the known transition-metal derivatives containing both π-cyclopentadienyl and carbonyl ligands. In both cases, compounds with the eighteen-electron rare-gas electronic configuration are the most stable. However, the relatively stable but oxygen-sensitive biscyclopentadienyls of the first-row transition metals $(C_5H_5)_2M$ (M = V, Cr, Co, and Ni), do not have the rare-gas configuration. Among the cyclopentadienylmetal carbonyls (Table I-9) only a few polynuclear ones, such as $(C_5H_5)_2V_2(CO)_5$ and $(C_5H_5)_3Ni_3(CO)_2$, do not have the favored eighteen-electron rare-gas electronic configuration.

PREPARATIVE METHODS

The most generally applicable method for the preparation of metal π-cyclo-pentadienyl derivatives is the reaction between sodium cyclopentadienide and

TABLE I-8

Some Metal Cyclopentadienyls[a]

Group	III	IV	V	VI	VII	VIII	VIII	VIII	IB
Electrons needed to attain rare-gas configuration	15	14	13	12	11	10	9	8	7
3d Transition series		$[Cp_2Ti]_2$ green	Cp_2V violet	Cp_2Cr^+ yellow Cp_2Cr red	Mn^{2+}	Cp_2Fe^+ blue Cp_2Fe orange	Cp_2Co^+ yellow Cp_2Co purple	Cp_2Ni^+ yellow Cp_2Ni green	
4d Transition series		$[Cp_2Zr]_n$ purple		Cp_2MoH_2 yellow	Cp_2TcH yellow	Cp_2Ru yellow	Cp_2Rh^+ colorless (Cp_2Rh, Cp_4Rh_2) yellow Cp_4Rh_3H black		
5d Transition series	Lanthanide^{3+}, Eu^{2+}, Yb^{2+}		Cp_2TaH_3 white	Cp_2WH_2 yellow	Cp_2ReH yellow	Cp_2Os colorless	Cp_2Ir^+ colorless (Cp_2Ir, Cp_4Ir_2) yellow	Cp_4Pt_2 dark green	$[CpAu]_n$ yellow

[a] $Cp = \pi$-cyclopentadienyl. Listing of a simple ion means that an ionic cyclopentadienide of this ion has been prepared.

TABLE I-9

SOME CYCLOPENTADIENYLMETAL CARBONYLS[a]

Group	III	IV	V	VI	VII	VIII	VIII	VIII	IB
Electrons needed to attain rare-gas configuration	15	14	13	12	11	10	9	8	7
3d Transition series		$Cp_2Ti(CO)_2$ red-brown cryst.	$CpV(CO)_4$ orange cryst. $Cp_2V_2(CO)_5$ green cryst.	$[CpCr(CO)_3]_2$ green cryst.	$CpMn(CO)_3$ yellow cryst.	$[CpFe(CO)_2]_2$ red-violet cryst. $[CpFeCO]_4$ green cryst.	$CpCo(CO)_2$ red liquid $[CpCoCO]_3$ black cryst.	$[CpNiCO]_2$ red-green cryst. $Cp_3Ni_3(CO)_2$ green-brown cryst.	
4d Transition series			$CpNb(CO)_4$ orange cryst.	$[CpMo(CO)_3]_2$ red-violet cryst.	$CpTc(CO)_3$ white cryst.	$[CpRu(CO)_2]_2$ yellow-orange cryst. $[CpRuCO]_4$ red-violet cryst.	$CpRh(CO)_2$ orange liq. $Cp_2Rh_2(CO)_3$ red cryst. $[CpRhCO]_3$ black cryst.		
5d Transition series			$CpTa(CO)_4$ orange cryst.	$[CpW(CO)_3]_2$ red-violet cryst.	$CpRe(CO)_3$ white cryst.	$[CpOs(CO)_2]_2$ yellow-orange cryst.	$CpIr(CO)_2$ yellow liq.	$[CpPtCO]_2$ red-violet cryst.	

[a] $Cp = \pi$-cyclopentadienyl.

an appropriate transition-metal halide [23]. Sodium cyclopentadienide may be readily obtained by reaction of metallic sodium or sodium hydride with cyclopentadiene in a polar solvent such as tetrahydrofuran, 1,2-dimethoxy-ethane, diglyme, diethyl ether, t-butanol, or liquid ammonia. Tetrahydrofuran (THF) is most frequently used. Sodium cyclopentadienide is a pyrophoric white solid, but its solutions can be handled satisfactorily under an inert atmosphere such as nitrogen or argon.

The following equations illustrate applications of sodium cyclopentadienide for the preparation of various π-cyclopentadienyl derivatives.

$$2\ NaC_5H_5 + CoCl_2 \xrightarrow{\ THF\ } (C_5H_5)_2Co + 2\ NaCl \qquad (a)$$

$$[(C_2H_4)_2RhCl]_2 + 2\ NaC_5H_5 \xrightarrow{\ THF\ } 2\ C_5H_5Rh(C_2H_4)_2 + 2\ NaCl \qquad (b)$$

$$2\ NaC_5H_5 + TiCl_4 \xrightarrow{\ THF\ } (C_5H_5)_2TiCl_2 + 2\ NaCl \qquad (c)$$

Sometimes other ionic cyclopentadienides are used instead of sodium cyclopentadienide for the introduction of the π-cyclopentadienyl ligand into transition-metal complexes. These alternative reagents may include other alkali-metal cyclopentadienides such as potassium cyclopentadienide (from potassium and cyclopentadiene) or lithium cyclopentadienide (from an alkyllithium compound and cyclopentadiene). Magnesium cyclopentadienide, a volatile pyrophoric solid, may also be used for the introduction of the π-cyclopentadienyl ligand in some cases [24]. This magnesium derivative is readily obtained by the high-temperature reaction of magnesium metal with cyclopentadiene. An air-stable cyclopentadienide which may be used for the preparation of some π-cyclopentadienyl derivatives is thallium cyclopenta-dienide, a very pale yellow, volatile solid which may be prepared in good yield by addition of cyclopentadiene to strongly basic aqueous solutions of thallium(I) salts [25].

In the more favorable cases the metal cyclopentadienide normally required for the preparation of a π-cyclopentadienylmetal derivative can be replaced with a mixture of free cyclopentadiene and a strong base such as an amine, sodium alkoxide, or a very concentrated aqueous solution of an alkali-metal hydroxide. Apparently, in the cases of the more readily formed π-cyclopenta-dienylmetal derivatives, the very small equilibrium concentrations of the cyclopentadienide anion in the mixture of cyclopentadiene and a strong base are sufficient to form the π-cyclopentadienylmetal derivative. Amines most often serve the function of generating equilibrium quantities of the cyclo-pentadienide anion from cyclopentadiene, as illustrated by the following preparation of nickelocene [26]:

$$NiBr_2 + 2\ Et_2NH + 2\ C_5H_6 \longrightarrow (C_5H_5)_2Ni + 2\ [Et_2NH_2]Br$$

Mercury derivatives are frequently used in preparative organometallic chemistry but are infrequently used in the preparation of π-cyclopentadienyl-metal derivatives. A solution apparently containing cyclopentadienylmercuric chloride can be obtained by reacting sodium cyclopentadienide and mercuric chloride in a 1:1 molar ratio. This mercury derivative reacts with the hexa-carbonylmetallates of vanadium, niobium, and tantalum to give the corresponding cyclopentadienylmetal tetracarbonyls [13b]:

$$M(CO)_6^- + C_5H_5HgCl \longrightarrow C_5H_5M(CO)_4 + 2\ CO + Hg + Cl^-$$

where M = vanadium, niobium, or tantalum.

Metal carbonyls are often useful intermediates for the preparation of π-cyclopentadienyl derivatives. Thus tris(acetonitrile)tungsten tricarbonyl reacts with cyclopentadiene in boiling hexane to give $C_5H_5W(CO)_3H$ according to the following equation [27]:

$$(CH_3CN)_3W(CO)_3 + C_5H_6 \xrightarrow[80°]{\text{hexane}} C_5H_5W(CO)_3H + 3\ CH_3CN$$

One of the hydrogens of the CH_2 group in cyclopentadiene is transferred from the carbon atom to the tungsten. A similar metal carbonyl reaction with cyclopentadiene provides useful methods for preparing $[C_5H_5Fe(CO)_2]_2$ and $C_5H_5Co(CO)_2$ as follows [28]:

$$2\ Fe(CO)_5 + 2\ C_5H_6 \xrightarrow[130°]{\text{dicyclopentadiene}} [C_5H_5Fe(CO)_2]_2 + 6\ CO + [H_2]$$

$$Co_2(CO)_8 + 2\ C_5H_6 \xrightarrow[25°]{CH_2Cl_2} 2\ C_5H_5Co(CO)_2 + 4\ CO + [H_2]$$

In these cases, the cyclopentadienylmetal carbonyl hydride of iron or cobalt corresponding to $C_5H_5W(CO)_3H$ is so unstable that it liberates hydrogen under the reaction conditions to give the hydride-free cyclopentadienylmetal carbonyl. Related reactions are those of sodium cyclopentadienide with the hexacarbonyls of chromium, molybdenum, or tungsten in boiling tetrahydro-furan, 1,2-dimethoxyethane, diglyme, or dimethylformamide to give the anions $C_5H_5M(CO)_3^-$ (M = Cr, Mo, or W) according to the following equation [29]:

$$M(CO)_6 + NaC_5H_5 \xrightarrow{\Delta} Na[C_5H_5M(CO)_3] + 3\ CO$$

where M = chromium, molybdenum, or tungsten.

Metal Complexes of Other C_nH_n Ring Systems and Their Derivatives

In addition to the cyclopentadienyl ring (C_5H_5), other related planar polygonal C_nH_n ring systems and their derivatives can form π-bonds to transition metals which use all n carbon atoms in the bonding. Other ring systems of this type which form a variety of π-complexes are cyclobutadiene (C_4H_4), benzene (C_6H_6), and cycloheptatrienyl (C_7H_7). Up to now, the planar octagonal C_8H_8 ring has only been found in the two outer rings of the single titanium complex $(C_8H_8)_3Ti_2$ [30] and in the uranium derivative $(C_8H_8)_2U$. No derivatives of the unsubstituted π-cyclopropenyl ring (C_3H_3) are known, but the substituted π-triphenylcyclopropenyl ring is found in a very few complexes such as $(C_6H_5)_3C_3Ni(CO)Br$ [31]. In general, the C_4H_4, C_6H_6, and C_7H_7 complexes are less stable than analogous C_5H_5 complexes.

The bonding of transition metals to these other C_nH_n rings follows the general pattern outlined above (Fig. I-2) for the metal–π-cyclopentadienyl bond. In the cases of the C_4H_4, C_6H_6, and C_7H_7 metal complexes, as in the case of the C_5H_5 complexes discussed above, the forward portion of the metal–ring bond is a triple bond with one σ-bonding component and two orthogonal π-bonding components. Retrodative δ-bonding also occurs.

PREPARATION OF π-C_4H_4 COMPLEXES

The unavailability of free cyclobutadiene makes the preparation of cyclobutadiene–metal complexes relatively difficult. However, the dehalogenation of 3,4-dichlorocyclobutene ($C_4H_4Cl_2$) with various metal carbonyl derivatives provides a suitable route to several cyclobutadienemetal complexes such as the following [32, 33]:

$$Fe_2(CO)_9 + C_4H_4Cl_2 \xrightarrow[25°]{ether} C_4H_4Fe(CO)_3 + 6\ CO + FeCl_2$$

$$Na_2Ru(CO)_4 + C_4H_4Cl_2 \longrightarrow C_4H_4Ru(CO)_3 + 2\ NaCl + CO$$

$$Na_2M(CO)_5 + C_4H_4Cl_2 \longrightarrow C_4H_4M(CO)_4 + 2\ NaCl + CO$$

where M = Cr, Mo, and W. The cyclobutadiene derivatives $C_4H_4Fe(CO)_3$ and $C_5H_5CoC_4H_4$ can also be obtained by ultraviolet irradiation of α-pyrone (I) with $Fe(CO)_5$ or $C_5H_5Co(CO)_2$, respectively; carbon dioxide is eliminated in this reaction [34].

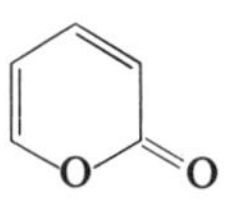

I

A larger variety of metal derivatives of tetrasubstituted cyclobutadienes have been prepared. The first cyclobutadiene derivative to be prepared was tetramethylcyclobutadienenickel dichloride reported in 1959 by Criegee and Schröder as being formed in good yield in the following reaction [32]:

$$2\ (CH_3)_4C_4Cl_2 + 2\ Ni(CO)_4 \longrightarrow [(CH_3)_4C_4NiCl_2]_2 + 8\ CO$$

Tetramethylcyclobutadiene derivatives of other metals can be prepared from this nickel complex by a cyclobutadiene transfer reaction [35]:

$$2\ Co_2(CO)_8 + [(CH_3)_4C_4NiCl_2]_2 \longrightarrow 2(CH_3)_4C_4Co(CO)_2Co(CO)_4 + 2\ CO + 2\ NiCl_2$$

Metal complexes of tetraphenylcyclobutadiene can sometimes be prepared by treatment of diphenylacetylene with an appropriate metal derivative:

$$Fe(CO)_5 + 2\ C_6H_5C_2C_6H_5 \xrightarrow{\ \Delta\ } (C_6H_5)_4C_4Fe(CO)_3 + 2\ CO \qquad (a)$$

$$2\ PdCl_2 + 4\ C_6H_5C_2C_6H_5 \xrightarrow{\ HCl\ } [(C_6H_5)_4C_4PdCl_2]_2 \qquad (b)$$

The palladium complex $[(C_6H_5)_4C_4PdCl_2]_2$ can be used to prepare tetraphenyl-cyclobutadiene derivatives of other metals by means of cyclobutadiene transfer reactions [36]:

$$Co_2(CO)_8 + [(C_6H_5)_4C_4PdCl_2]_2 \longrightarrow 2(C_6H_5)_4C_4Co(CO)_2Cl + 4\ CO + Pd + PdCl_2 \quad (a)$$

$$[C_5H_5Mo(CO)_3]_2 + [(C_6H_5)_4C_4PdCl_2]_2 \longrightarrow$$
$$2\ C_5H_5Mo(CO)ClC_4(C_6H_5)_4 + 4\ CO + Pd + PdCl_2 \quad (b)$$

Preparation of π-Complexes of Benzene and Related Aromatic Compounds

Some π-complexes of benzene and related aromatic compounds can be prepared by heating the parent ligand with certain metal carbonyl derivatives. Particularly useful are the reactions between hexacarbonylchromium and benzenoid compounds (arenes) to give arene–chromium tricarbonyls according to the following equation [37]:

$$Arene + Cr(CO)_6 \xrightarrow{110°-250°} (arene)Cr(CO)_3 + 3\ CO$$

The large variety of arenes reacting with $Cr(CO)_6$ in this manner to form $(arene)Cr(CO)_3$ complexes includes benzene and all its methylated derivatives, aniline and other similar amines, methyl benzoate, chlorobenzene, anisole, biphenyl, naphthalene, phenanthrene, and many others. The derivatives of benzoic acid and benzaldehyde cannot be prepared directly but can be obtained indirectly by hydrolysis of the appropriate esters and acetals, respectively. Arenes with many electronegative substituents, such as several halogen atoms, do not form arene–chromium tricarbonyl complexes, since the bonding orbitals of the benzene ring no longer have sufficient electron density to form stable forward bonds to the chromium atom.

Several metal carbonyls besides $Cr(CO)_6$ react directly with certain arenes to form arene–metal carbonyl derivatives. The hexacarbonyls of molybdenum and tungsten react, like $Cr(CO)_6$, with certain arenes to form the corresponding arene–metal tricarbonyls [37]. The acetonitrile complex

$(CH_3CN)_3W(CO)_3$ forms $(arene)W(CO)_3$ complexes upon heating with benzene, toluene, p-xylene, or mesitylene under milder conditions than hexacarbonyltungsten [27]. Hexacarbonylvanadium reacts with benzene and its methylated derivatives to form ionic products according to the following equation [38]:

$$2\ V(CO)_6 + arene \xrightarrow{50°} [(arene)V(CO)_4]\,[V(CO)_6] + 2\ CO$$

Dodecacarbonyltriruthenium $Ru_3(CO)_{12}$ reacts with benzene and its methylated derivatives to form hexametallic carbonyl carbide derivatives of the general formula $(arene)Ru_6(CO)_{14}C$ [9].

Sometimes the reactions of metal carbonyl derivatives with arenes require an aluminum halide catalyst in order to obtain arene–metal carbonyl complexes. This is particularly true in the preparations of arene–metal carbonyl cations from metal carbonyl halide derivatives, as exemplified by the following three reactions, all carried out in boiling benzene in the presence of aluminum chloride [39]:

$$Mn(CO)_5Br + C_6H_6 \xrightarrow[\text{AlCl}_3]{80°} [C_6H_6Mn(CO)_3]^+ + Br^- + 2\ CO \tag{a}$$

$$C_5H_5Fe(CO)_2Cl + C_6H_6 \xrightarrow[\text{AlCl}_3]{80°} [C_5H_5FeC_6H_6]^+ + Cl^- + 2\ CO \tag{b}$$

$$C_5H_5M(CO)_3Cl + C_6H_6 \xrightarrow[\text{AlCl}_3]{80°} [C_5H_5M(CO)C_6H_6]^+ + Cl^- + 2\ CO \tag{c}$$

where $M = Mo$ or W. The $[C_5H_5FeC_6H_6]^+$ cation can be prepared more conveniently by reaction of a mixture of ferrocene, aluminum chloride, and aluminum powder in boiling benzene solution. This reaction substitutes a benzene ring for one of the π-cyclopentadienyl rings of ferrocene. All of these arene–metal cations can be conveniently precipitated as hexafluorophosphate salts by addition of ammonium hexafluorophosphate to their aqueous solutions.

A standard method for the preparation of diarene–metal complexes utilizes the reaction of an anhydrous metal halide with the aromatic compound at elevated temperatures in the presence of excesses of aluminum powder as a reducing agent and an anhydrous aluminum halide as a catalyst. This reaction, which is called the reducing Friedel–Crafts reaction, is exemplified by the following equations [4, 40]:

$$VCl_4 + 2\ C_6H_6 + Al \xrightarrow{\text{AlCl}_3} [(C_6H_6)_2V]\,[AlCl_4] \tag{a}$$

$$3\ CrCl_3 + 6\ C_6H_6 + 2\ Al + AlCl_3 \longrightarrow 3\ [(C_6H_6)_2Cr]\,[AlCl_4] \tag{b}$$

$$3\ MCl_5 + 6\ C_6H_6 + 4\ Al \xrightarrow{\text{AlCl}_3} 3\ [(C_6H_6)_2M]\,[AlCl_4] + AlCl_3 \tag{c}$$

$$FeBr_2 + 2\ C_6H_3Me_3 + 2\ AlBr_3 \longrightarrow [(C_6H_3Me_3)_2Fe]\,[AlBr_4]_2 \tag{d}$$

$$3\ RuCl_3 + 6\ C_6H_6 + Al + 5\ AlCl_3 \longrightarrow 3\ [(C_6H_6)_2Ru]\,[AlCl_4]_2 \tag{e}$$

where M = Mo or Re in Eq. (c). The addition of the aluminum powder can be omitted in cases where the oxidation states of the metal atom in the diarene–metal cation and the starting metal halides are the same [e.g., Eq. (d) above]. The diarene–metal cations may be isolated as their sparingly water-soluble hexafluorophosphate salts. Furthermore, the diarene–metal cations of vanadium, chromium, molybdenum, and tungsten can be converted to the neutral $(C_6H_6)_2$ M complexes (M = V, Cr, Mo, or W) in strong aqueous base, sometimes with added reducing agent, generally sodium dithionite $Na_2S_2O_4$ [40].

All of these preparative methods for π-arene complexes use the arene itself as the source of the π-complexed arene ligand. Other preparative methods use arylmagnesium halides, acetylene trimerization, or cyclohexadiene dehydrogenation as the source of the π-arene ligand. The following illustrates the uses of some of these alternate means for introducing π-arene ligands.

Arylmagnesium Halides

The first π-arene derivatives to be prepared (but not recognized as such) were the "polyphenyl derivatives of chromium" obtained by Hein and co-workers around 1920 from the reaction between phenylmagnesium bromide and anhydrous chromium(III) chloride followed by hydrolysis [41]. The rather complex reaction mixture obtained by this procedure contains the diarene–chromium complexes of benzene and biphenyl as well as a mixed benzene–biphenyl–chromium complex. Much more recently, reactions between phenylmagnesium halides, sodium cyclopentadienide, and appropriate metal halides have been used to prepare the "mixed sandwich" compounds $C_5H_5MC_6H_6$ (M = Cr or Mn) [42].

Acetylene Trimerizations

This synthetic method is used but rarely, and then primarily for preparations of π-complexes of completely substituted arenes such as hexamethylbenzene. Thus, chromium halide derivatives react with an organomagnesium or organoaluminum compound in the presence of butyne-2 (dimethylacetylene) to give a bis(hexamethylbenzene)chromium complex [43].

Cyclohexadiene Dehydrogenation [44]

In some cases a π-benzene ligand can arise from dehydrogenation and complex formation of cyclohexadiene, generally the 1,3-isomer. For example, reduction of a mixture of chromium(III) chloride and 1,3-cyclohexadiene with isopropylmagnesium bromide gives dibenzene–chromium, $(C_6H_6)_2Cr$. An analogous reaction but using anhydrous iron(III) chloride instead of anhydrous chromium(III) chloride gives $C_6H_6FeC_6H_8$ (II), a mixed π-complex of both benzene and cyclohexadiene.

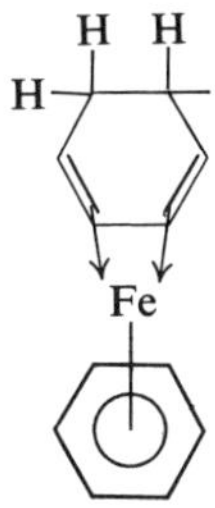

II

PREPARATION OF CYCLOHEPTATRIENYL DERIVATIVES

Relatively few π-cycloheptatrienyl derivatives are known both because of their lower stability as compared with analogous π-benzene and π-cyclopentadienyl derivatives, and because of the availability of fewer and less general synthetic methods. Some π-cycloheptatrienyl derivatives can be obtained by treating various metal carbonyl complexes with cycloheptatriene, the C_7H_8 hydrocarbon being converted to a π-C_7H_7 group with loss of a hydrogen atom. Reactions of this type seem to occur most readily with vanadium carbonyl derivatives. Thus, $C_5H_5V(CO)_4$ reacts with boiling cycloheptatriene to give the purple, mixed sandwich compound $C_5H_5VC_7H_7$ [45]. Similarly, $V(CO)_6$ reacts with cycloheptatriene to give a mixture of the green, nonionic π-cycloheptatrienyl derivative $C_7H_7V(CO)_3$ and the brown, ionic, mixed π-cycloheptatrienyl–π-cycloheptatriene derivative $[C_7H_7VC_7H_8][V(CO)_6]$ [46]. In other cases, π-cycloheptatrienyl derivatives can be obtained by hydride abstraction from corresponding π-cycloheptatriene–metal derivatives with triphenylmethyl salts in dichloromethane or acetonitrile solution at room temperature [47]:

$$C_7H_8M(CO)_3 + [(C_6H_5)_3C][BF_4] \xrightarrow{CH_2Cl_2} [C_7H_7M(CO)_3][BF_4] \downarrow + (C_6H_5)_3CH$$

where M = Cr, Mo, and W. Tropylium bromide is but rarely a useful reagent for the preparation of π-cycloheptatrienyl derivatives, presumably owing to repulsion between the positively charged tropylium ion and the positively charged metal atom; however, the cation $[C_5H_5CrC_7H_7]^+$ can be synthesized from $C_5H_5CrC_6H_6$ and tropylium salts.

Metal–Olefin Complexes

BONDING

Another frequently encountered ligand type in transition-metal organometallic chemistry includes olefins, diolefins, triolefins, and related species with carbon–carbon double bonds. In order to clarify the nature of the metal–

ligand bond in these complexes, the mode of bonding of transition metals to ethylene, the simplest olefin, will be considered. The salient features of the metal–olefin bond are depicted in Fig. 3. This figure considers the molecular orbitals of the carbon–carbon π-bond, since these are the olefin molecular orbitals of importance in forming a metal–olefin bond.

The carbon–carbon π-bond has two molecular orbitals: a filled bonding orbital with no nodes and an empty antibonding orbital with one node. The

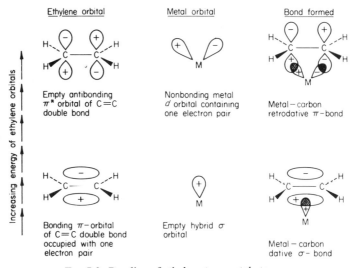

Ethylene orbital   Metal orbital   Bond formed

Increasing energy of ethylene orbitals

Empty antibonding π^* orbital of C=C double bond

Nonbonding metal d orbital containing one electron pair

Metal–carbon retrodative π–bond

Bonding π–orbital of C=C double bond occupied with one electron pair

Empty hybrid σ orbital

Metal–carbon dative σ–bond

FIG. I-3. Bonding of ethylene to a metal atom.

filled bonding orbital can overlap with an empty transition-metal hybrid orbital to form a σ-bond. In forming this σ-bond, the electron pair is donated to the metal atom through a dative bond. The empty antibonding orbital with one node has the appropriate symmetry to overlap with the filled metal d orbitals. This results in a retrodative bond between the metal atom and the olefinic ligand. Olefins thus resemble carbon monoxide in having available appropriate antibonding orbitals for retrodative π-bonding. However, each carbon–carbon double bond has only one antibonding orbital available for retrodative π-bonding, in contrast to carbon monoxide which has two orthogonal antibonding orbitals available for retrodative π-bonding. This difference may account for the observed greater retrodative π-bonding in metal carbonyls than in metal–olefin complexes.

TYPES OF COMPOUNDS FORMED

The variety of monoolefins forming transition-metal complexes is wide, ranging from unsubstituted ethylene to derivatives with electronegative substituents such as acrylonitrile, acrylic acid derivatives, maleic anhydride, maleic and fumaric acid derivatives, and many others. Even such unusual olefins as tetrafluoroethylene and tetracyanoethylene form metal complexes. The diolefins forming transition-metal complexes include both conjugated ones such as butadiene and 1,3-cyclohexadiene, and nonconjugated ones such

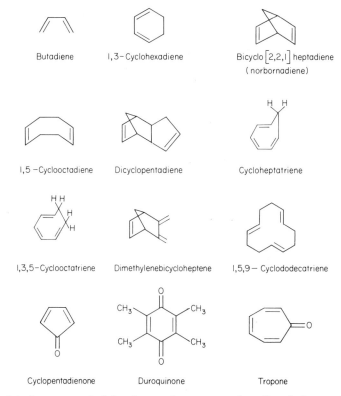

Butadiene   1,3-Cyclohexadiene   Bicyclo[2,2,1] heptadiene
(norbornadiene)

1,5-Cyclooctadiene   Dicyclopentadiene   Cycloheptatriene

1,3,5-Cyclooctatriene   Dimethylenebicycloheptene   1,5,9-Cyclododecatriene

Cyclopentadienone   Duroquinone   Tropone

FIG. I-4. Structures of olefins frequently encountered as ligands in transition-metal organometallic chemistry.

as norbornadiene, 1,5-cyclooctadiene, and dicyclopentadiene. Subtle changes in the relative locations of the double bonds in diolefins often greatly influence their complexing behavior with transition metals. The triolefins forming transition-metal complexes include cycloheptatriene, dimethylenebicyclo-heptene, and 1,3,5-cyclooctatriene. In addition, unsaturated ketones of both the cyclic type, such as cyclopentadienone or duroquinone, and the open-chain type, such as methyl vinyl ketone, form some complexes with transition metals. Structures of some of the most important olefinic ligands in transition-metal organometallic chemistry are depicted in Fig. I-4.

PREPARATIVE METHODS

In almost all cases the preparation of metal–olefin complexes simply involves reaction of the olefin with an appropriate transition-metal derivative. The types of transition-metal derivatives used and the reaction conditions for different systems are the two points of particular concern in this area of preparative transition-metal organometallic chemistry. The following examples illustrate some of the more frequently encountered systems for preparing transition metal–olefin complexes.

Monoolefin Complexes

Preparations of ethylene complexes will be used to illustrate some of the syntheses of monoolefin–transition-metal complexes. Many of these reactions can be extended to other monoolefins.

$$RhCl_3(H_2O)_3 + C_2H_4 \xrightarrow[H_2O]{ROH} [(C_2H_4)_2RhCl]_2 + \text{other products}$$

$$C_5H_5Mn(CO)_3 + C_2H_4 \xrightarrow{UV} C_5H_5Mn(CO)_2C_2H_4 + CO$$

$$Fe_2(CO)_9 + C_2H_4 \xrightarrow{25°/50 \text{ atm}} C_2H_4Fe(CO)_4 + Fe(CO)_5$$

Conjugated Diolefin Complexes

The systems depicted below represent cases where conjugated diolefins behave differently from nonconjugated diolefins.

$$Fe(CO)_5 + C_4H_6 \xrightarrow{130°} C_4H_6Fe(CO)_3 + 2 CO$$

$$2 (1,3\text{-}C_6H_8) + (CH_3CN)_3M(CO)_3 \xrightarrow{80°} (C_6H_8)_2M(CO)_2 + CO + 3 CH_3CN$$

(M = Mo and W)

$$IrCl_3 + 1,3\text{-}C_6H_8 \xrightarrow{EtOH} (C_6H_8)_2IrCl + \text{other products}$$

Nonconjugated Diolefin Complexes

Preparations of 1,5-cyclooctadiene complexes will be used to illustrate some of the commonly used preparations of metal complexes of nonconjugated diolefins.

$$C_5H_5Co(CO)_2 + 1,5\text{-}C_8H_{12} \xrightarrow{135°} C_5H_5CoC_8H_{12} + 2 CO$$

$$Mo(CO)_6 + 1,5\text{-}C_8H_{12} \xrightarrow{100°} C_8H_{12}Mo(CO)_4 + 2 CO$$

$$RhCl_3(H_2O)_3 + 1,5\text{-}C_8H_{12} \xrightarrow[80°]{ethanol} [C_8H_{12}RhCl]_2 + \text{other products}$$

$$(CH_3)_4C_6O_2 + Ni(CO)_4 + 1,5\text{-}C_8H_{12} \longrightarrow [(CH_3)_4C_6O_2]NiC_8H_{12} + 4 CO$$

Duroquinone

Unsaturated Ketone Complexes

$$2 \ CF_3C\equiv CCF_3 + Fe(CO)_5 \longrightarrow (CF_3)_4C_5OFe(CO)_3 + CO$$

$$[(CF_3)_4C_5O = \text{tetrakis(trifluoromethyl)cyclopentadienone}]$$

$$3 \ CH_3COCH\!=\!CH_2 + (CH_3CN)_3W(CO)_3 \xrightarrow[80°]{\text{hexane}} (CH_3COCH\!=\!CH_2)_3W + 3 \ CH_3CN + 3 \ CO$$

These olefin complexes will be discussed in greater detail in later sections dealing with the specific metals involved.

Metal–Alkyne (Acetylene) Complexes

BONDING

The carbon–carbon triple bond of acetylene and its derivatives consists of a single carbon–carbon σ-bond flanked with two perpendicular carbon–carbon π-bonds. This system of bonds can form complexes with metals in the following three ways (Fig. I-5).

Monodentate Monometallic

One of the carbon–carbon π-bonds of the alkyne is bonded to the metal atom just like the carbon–carbon π-bond of an olefin. The forward dative

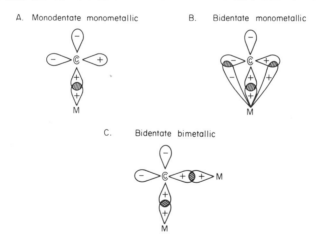

FIG. I-5. Types of bonding between alkynes and metal atoms. For clarity, the possibilities of retrodative bonding (both π and δ) are ignored in this figure. The retrodative bonding possibilities resemble those in metal–olefin and metal–π-cyclopentadienyl complexes (Figs. I-2 and I-3). The acetylene ligand is viewed end-on (i.e., with the C—C axis perpendicular to the paper) and only the π-orbitals are shown.

σ-bond and the retrodative π-bond of a monodentate monometallic alkyne complex are very similar to those found in the olefin complexes discussed above. The second carbon–carbon π-bond of the alkyne is not involved in the bonding to the metal atom.

Bidentate Monometallic

One of the carbon–carbon π-bonds of the alkyne bonds to a metal atom, just like the carbon–carbon π-bond of an olefin, and just like the monodentate monometallic alkyne ligand. The second perpendicular carbon–carbon π-bond of the alkyne forms a forward π-bond with empty p or d orbitals of the metal atom, just as the E_1 orbitals of the C_5H_5 ring form a metal π-cyclopentadienyl bond. In addition, the antibonding orbitals of this second carbon–carbon π-bond can form a retrodative δ-bond with appropriate d orbitals of the metal atom.

Bidentate Bimetallic

One of the carbon–carbon π-bonds of the alkyne is bonded to a metal atom like the carbon–carbon π-bond of an olefin and the monodentate monometallic alkyne ligand. The second carbon–carbon π-bond of the alkyne is similarly bonded to a second metal atom. Most such bidentate bimetallic alkyne–metal complexes have a metal–metal bond between the two metal atoms.

TYPES OF COMPOUNDS FORMED

Most reactions of metal complexes with alkynes do not yield simple alkyne complexes of any of the three types cited above. Instead, in most cases, two or more molecules of the introduced alkyne combine to form derivatives of cyclobutadiene, benzene, or other more complex organic systems. Sometimes the metal atom may form a heterocyclic ring system with two or three alkyne fragments, giving metallacyclopentadiene or metallacycloheptatriene derivatives. If metal carbonyls are involved, insertion of carbon monoxide may accompany oligomerization of the alkyne, resulting in derivatives of cyclic unsaturated ketones such as cyclopentadienones, quinones, or tropones. Thus, formation of actual alkyne–metal complexes from reactions of alkynes with metal carbonyls or other complexes occurs only in a relatively small minority of the cases because of competition with oligomerization reactions of the alkyne.

PREPARATIVE METHODS

The following reactions are representative of the minority of cases where true alkyne–metal complexes rather than metal derivatives of alkyne oligomerization products are formed in reactions of metal derivatives with alkynes.

i. *Monodentate monometallic derivatives of alkynes*

$$C_5H_5Mn(CO)_3 + RC\equiv CR \xrightarrow{UV} C_5H_5Mn(CO)_2C_2R_2 + CO$$

ii. *Bidentate monometallic derivatives of alkynes [48]*

$$C_5H_5V(CO)_4 + RC\equiv CR \xrightarrow{UV} C_5H_5V(CO)_2C_2R_2 + 2\ CO$$

$$(CH_3CN)_3W(CO)_3 + 3\ RC\equiv CR \xrightarrow{80°} (R_2C_2)_3WCO + 2\ CO + 3\ CH_3CN$$

The alkyne ligand replaces two carbonyl groups in $C_5H_5V(CO)_4$ when it forms the bidentate monometallic complex $C_5H_5V(CO)_2C_2R_2$. However, the alkyne ligand replaces only one carbonyl group in $C_5H_5Mn(CO)_3$ when it forms the monodentate monometallic complex $C_5H_5Mn(CO)_2C_2R_2$.

iii. *Bidentate bimetallic derivatives of alkynes [49]*

$$Co_2(CO)_8 + RC\equiv CR \xrightarrow{25°} R_2C_2Co_2(CO)_6 + 2\ CO$$

$$[C_5H_5NiCO]_2 + RC\equiv CR \xrightarrow{65°} R_2C_2(NiC_5H_5)_2 + 2\ CO$$

Metal π-Allyl Derivatives

Bonding

The π-allyl ligand may be regarded as an open-chain fragment corresponding to half of a benzene ring. Figure I-6 indicates the molecular orbitals of the π-allyl ligand and its mode of bonding with a metal atom.

If the π-allyl ligand is regarded as the anion $C_3H_5^-$, it may be considered as a bidentate ligand which donates two electron pairs to the metal atom by means of a double bond with the following two components: (1) a σ-bond involving overlap of the filled bonding B_1 orbital with an empty hybrid orbital of the metal atom; (2) a π-bond involving overlap of the filled nonbonding A_2 orbital with empty p or d orbitals of the metal atom of the same symmetry.

The double bond between the metal atom and the π-allyl ligand thus consists of one σ-bond and one π-bond analogous to the carbon–carbon double bond of ethylene. In addition, some features of the π-allyl–metal double bond resemble those of the π-cyclopentadienyl–metal triple bond. A further component of the π-allyl-metal bond is partial retrodative δ-bonding of the filled metal d orbitals with the empty π-allyl antibonding B_1 orbital; this resembles the corresponding retrodative δ-bonding in π-cyclopentadienyl derivatives.

Types of Compounds Formed

Table I-10 lists the known π-allylmetal derivatives containing no other ligands (conveniently designated as "isoleptic" π-allyl derivatives) [50]. In addition, allylmetal carbonyl derivatives of the types $C_3H_5Mn(CO)_4$, $C_3H_5Fe(CO)_3X$ (X = halogen, etc.), and $C_3H_5Co(CO)_3$ are known, as well as some of their derivatives. In general, π-allylmetal derivatives are much less stable than similar π-cyclopentadienyl metal derivatives.

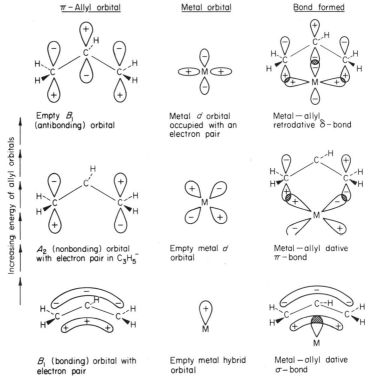

FIG. I-6. Bonding of a π-allyl group to a metal atom.

PREPARATIVE METHODS

Allyl halides are always the ultimate source of the π-C_3H_5 ligand in the preparation of π-allylmetal derivatives. Sometimes reaction of allyl halides with metal carbonyls or metal carbonyl anions will lead to the formation of π-allyl derivatives, as indicated by the following examples [51]:

$$Fe(CO)_5 + C_3H_5X \xrightarrow{50°} C_3H_5Fe(CO)_3X + 2\ CO$$

$$(X = Cl,\ Br,\ or\ I)$$

$$NaCo(CO)_4 + C_3H_5X \xrightarrow{25°} C_3H_5Co(CO)_3 + CO$$

$$[Rh(CO)_2Cl]_2 + C_3H_5Cl \longrightarrow [(C_3H_5)_2RhCl]_2 + other\ products$$

TABLE I-10

SOME ISOLEPTIC π-ALLYL DERIVATIVES OF THE TRANSITION METALS

Group	III	IV	V	VI	VII	VIII	VIII	VIII	IB
Electrons needed to attain rare-gas configuration	15	14	13	12	11	10	9	8	7
3d Transition series			$(C_3H_5)_3V$ brown, dec. −30°	$[(C_3H_5)_2Cr]_2$ red-brown, dec. 70°–80° $(C_3H_5)_3Cr$ red, mp 70°–80°		$(C_3H_5)_3Fe$ orange, dec. −40°	$(C_3H_5)_3Co$ golden red, dec. −40°	$(C_3H_5)_2Ni$ yellow, mp 1°, dec. 20°	
4d Transition series		$(C_3H_5)_4Zr$ red, dec. 0°	$(C_3H_5)_4Nb$ green-black, dec. 0°	$[(C_3H_5)_2Mo]_2$ deep green, dec. 110° $(C_3H_5)_4Mo$ green, dec. 125°			$(C_3H_5)_3Rh$ pale yellow, mp 80°–85°	$(C_3H_5)_2Pd$ pale yellow, mp 30°	
5d Transition series		$(C_3H_5)_4Hf$ red		$(C_3H_5)_4W$ light brown, dec. 95°			$(C_3H_5)_3Ir$ white, dec. 65°	$(C_3H_5)_2Pt$ colorless, mp 44°	

Other π-allyl derivatives may be obtained by converting the allyl halide to an allylmagnesium halide by reaction with magnesium followed by reaction of the allylmagnesium halide with an appropriate metal halide, as indicated by the following examples [50, 51]:

$$ZrCl_4 + 4\,C_3H_5MgCl \longrightarrow (C_3H_5)_4Zr + 4\,MgCl_2$$

$$NiBr_2 + 2\,C_3H_5MgBr \longrightarrow (C_3H_5)_2Ni + 2\,MgBr_2$$

$$[(C_3H_5)_2RhCl]_2 + 2\,C_3H_5MgCl \longrightarrow 2\,(C_3H_5)_3Rh + 2\,MgCl_2$$

$$[C_8H_{12}RhCl]_2 + 2\,C_3H_5MgCl \longrightarrow 2\,C_3H_5RhC_8H_{12} + 2\,MgCl_2$$

Some substituted π-allylmetal derivatives can be prepared by treatment of appropriate metal complexes with conjugated diolefins. Thus, butadiene reacts with $HCo(CO)_4$ to give the crotyl compound $\pi\text{-}CH_3C_3H_4Co(CO)_3$; this was the first reported π-allylic complex [52]. Butadiene reacts with $(C_6H_5CN)_2PdCl_2$ to give the π-allylic derivative $[ClCH_2C_3H_4PdCl]_2$. 1,3-Cyclohexadiene (C_6H_8) reacts with palladium(II) chloride to give the π-cyclohexenyl derivative $[C_6H_9PdCl]_2$ with takeup of one hydrogen atom for each 1,3-cyclohexadiene introduced. However, 1,3-cycloheptadiene reacts with palladium(II) chloride, resulting in loss of a hydrogen atom from each ligand introduced and giving the π-cycloheptadienyl complex $[C_7H_9PdCl]_2$ [53].

Other Delocalized Carbon Ligands

A few other delocalized systems of sp^2 carbon atoms have been shown to form π-complexes with transition metals. The Y-shaped trimethylenemethane ligand and the bisallylene ligand are each presently known in only one iron carbonyl complex. However, in both cases, the stability of the iron carbonyl complexes is sufficiently high that a variety of additional related complexes should be discovered in the future.

PREPARATION OF TRIMETHYLENEMETHANE–IRON TRICARBONYL [54]

$$Fe_2(CO)_9 + CH_2{=}C(CH_2Cl)_2 \longrightarrow$$

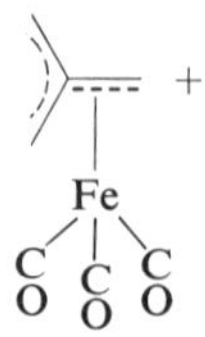

$+\ 6\,CO + FeCl_2$

PREPARATION OF BISALLYLENE–DIIRON HEXACARBONYL [55]

$$2 \, Fe_3(CO)_{12} + 6 \, H_2C{=}C{=}CH_2 \longrightarrow \qquad + 6 \, CO$$

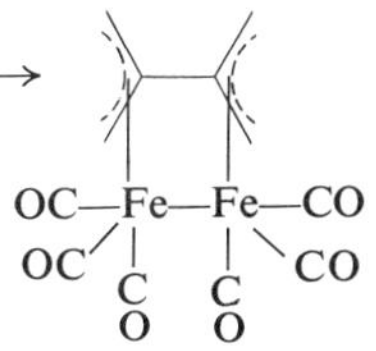

OC—Fe—Fe—CO
OC | | CO
     C   C
     O   O

Compounds with Transition-Metal–Carbon σ-Bonds

TYPES OF COMPOUNDS FORMED

The characteristic feature of the organometallic chemistry of the non-transition metals is the presence of a metal–carbon σ-bond. Most normal transition-metal organometallic derivatives with just metal–carbon σ-bonds (e.g., trimethyliron, diethylcobalt, hexaphenyltungsten, and tribenzylrhodium) analogous to nontransition-metal derivatives such as diethylzinc and triphenylaluminum are too unstable to exist under normal conditions [56]. In general the transition metals in such hypothetical derivatives with metal–carbon σ-bonds would have incompletely filled d orbitals often with unpaired electrons. In such cases metal–carbon σ-bonds appear to be very unstable with respect to homolytic cleavage, giving the free organic radical and the metal in a lower oxidation state. However, it is possible to stabilize compounds with transition-metal–carbon σ-bonds by attaching sufficient additional strong-field ligands to the metal atom to give it a stable electronic configuration and larger coordination number. Most often, such complexes have the favored

TABLE I-11

EXAMPLES OF STABILIZATION OF TRANSITION-METAL–CARBON σ-BONDS BY THE PRESENCE OF OTHER LIGANDS

Ligand	Derivatives stabilized by the ligand[a]
Carbonyl	$RMn(CO)_5$, $RCo(CO)_4$, $C_5H_5Fe(CO)_2R$, $C_5H_5Mo(CO)_3R$
Cyclopentadienyl	$(C_5H_5)_2TiR_2$, $(C_5H_5)_2VR$
Tertiary Phosphine	$(R_3P)_2PtR_2$, $(R'_3P)_2CoR_2$, $(diphos)_2RuR_2$
Cyanide	$[RCo(CN)_5]^{3-}$
Nitrosyl	$C_5H_5Cr(NO)_2R$

[a] R here refers to a group forming a metal–carbon σ-bond, such as alkyl, aryl, perfluoroalkyl, or perfluoroaryl.

rare gas configuration. Strong-field ligands suitable for stabilizing derivatives with transition-metal–carbon σ-bonds include π-acceptor organic ligands such as π-cyclopentadienyl and carbon monoxide. The compounds listed in Table I-11 exemplify derivatives with transition-metal–carbon σ-bonds which are stabilized by the presence of additional strong-field ligands, most often those with π-acceptor properties.

Compounds with a fluorocarbon group σ-bonded to a transition-metal atom are appreciably more stable than analogous compounds with a hydrocarbon group σ-bonded to a transition metal. The presence of fluorine on a carbon atom which is also bonded to a transition-metal atom stabilizes the metal–carbon bond but weakens the carbon–fluorine bond as a consequence of the contribution of no-bond resonance structures such as **IIIb** (for a trifluoro-methyl derivative). Since the metal–carbon σ-bond is generally the "weak link" in a transition-metal organometallic derivative, strengthening this bond stabilizes the compound. The weakened carbon–fluorine bond in fluorocarbon derivatives of transition metals relative to other fluorocarbon derivatives (because of resonance structures related to **IIIb**) is indicated by lower ν(CF) infrared frequencies, lower ^{19}F NMR chemical shifts of these fluorine atoms, and their easier removal by basic hydrolysis.

$$
\begin{array}{ccc}
\quad\ \text{F} & & \quad\ \text{F}^{\ominus} \\
\quad\ | & & \quad\ | \\
\text{F}\!-\!\text{C}\!-\!\text{M} & \longleftrightarrow & \text{F}\!-\!\text{C}\!=\!\text{M}^{\oplus} \\
\quad\ | & & \quad\ | \\
\quad\ \text{F} & & \quad\ \text{F}
\end{array}
$$

IIIa   **IIIb**

PREPARATIVE TECHNIQUES–NONFLUORINATED DERIVATIVES

Most preparations of compounds with transition-metal–carbon σ-bonds to nonfluorinated ligands utilize either the reaction of a reactive organometallic derivative of an electropositive metal such as sodium, lithium, magnesium, or aluminum with an appropriate transition-metal halide derivative, or the reaction between an alkyl halide and a low-valent transition-metal derivative such as a metal carbonyl anion. The following equations exemplify the use of reactive alkyl derivatives of lithium and magnesium for the preparation of compounds with transition-metal–carbon σ-bonds:

$$4\,CH_3Li + TiCl_4 \longrightarrow (CH_3)_4Ti + 4\,LiCl \tag{a}$$
(very
unstable)

$$(C_5H_5)_2TiCl_2 + 2\,LiC_6H_5 \longrightarrow (C_5H_5)_2Ti(C_6H_5)_2 + 2\,LiCl \tag{b}$$
(fairly stable)

$$6 \, C_6H_5MgBr + 2 \, CrCl_3 + 6 \, C_4H_8O \longrightarrow 2 \, (C_4H_8O)_3Cr(C_6H_5)_3 + 3 \, MgCl_2 + 3 \, MgBr_2$$

(sensitive to H₂

$$\text{(sensitive to } H_2 \quad \text{(c)}$$
$$\text{and } O_2)$$

$$2 \, C_5H_5Cr(NO)_2Cl + 2 \, CH_3MgI \longrightarrow 2 \, CH_3Cr(NO)_2C_5H_5 + MgCl_2 + MgI_2 \quad \text{(d)}$$
$$\text{(stable)}$$

$$(R_3P)_2NiBr_2 + 2 \, s\text{-}(CH_3)_3C_6H_2MgBr \longrightarrow (R_3P)_2Ni[C_6H_2(CH_3)_3]_2 + 2 \, MgBr_2 \quad \text{(e)}$$
$$\text{(stable with alkyl}$$
$$\text{group shown)}$$

$$4 \, K_2PtCl_6 + 12 \, CH_3MgI \longrightarrow [(CH_3)_3PtI]_4 + 8 \, MgCl_2 + 8 \, KCl + 4 \, MgI_2$$
$$\text{(stable)} \quad \text{(f)}$$

Similarly, the following equations exemplify the preparation of compounds with transition-metal–carbon σ-bonds using low-valent transition-metal intermediates such as metal carbonyl anions:

$$NaMn(CO)_5 + CH_3I \longrightarrow CH_3Mn(CO)_5 + NaI \quad \text{(a)}$$
$$\text{(stable)}$$

$$NaMn(CO)_5 + C_6H_5COCl \longrightarrow C_6H_5COMn(CO)_5 + NaCl \quad \text{(b)}$$
$$C_6H_5COMn(CO)_5 \overset{\Delta}{\longrightarrow} C_6H_5Mn(CO)_5 + CO$$

[Note: iodobenzene is too unreactive to give useful yields of $C_6H_5Mn(CO)_5$ from $NaMn(CO)_5$ in one step.]

$$K_6Co_2(CN)_{10} + CH_3I \longrightarrow K_3Co(CN)_5CH_3 + K_3Co(CN)_5I \quad \text{(c)}$$
$$\text{(stable)}$$

PREPARATIVE TECHNIQUES–FLUORINATED DERIVATIVES [57]

The changes in chemical properties of organic compounds upon replacement of a large percentage of the hydrogen with fluorine modify the methods for the preparation of fluorocarbon derivatives with transition metal–carbon σ-bonds. For example, the reaction of $NaMn(CO)_5$ with CF_3I does *not* give the trifluoromethyl derivative $CF_3Mn(CO)_5$; instead the iodide $Mn(CO)_5I$ is formed according to the following equation:

$$2 \, NaMn(CO)_5 + 2 \, CF_3I \longrightarrow 2 \, Mn(CO)_5I + 2 \, NaF + C_2F_4$$

However, the trifluoromethyl derivative $CF_3Mn(CO)_5$ can be prepared by the following sequence of reactions:

$$NaMn(CO)_5 + (CF_3CO)_2O \longrightarrow CF_3COMn(CO)_5 + CF_3CO_2Na$$

$$CF_3COMn(CO)_5 \overset{\Delta}{\longrightarrow} CF_3Mn(CO)_5 + CO$$

In other cases, however, fluorocarbon iodides react with neutral metal carbonyl derivatives to give transition metal–fluorocarbon products with metal–carbon σ-bonds, as exemplified by the following reactions:

$$Fe(CO)_5 + R_fI \xrightarrow{70°} R_fFe(CO)_4I + CO$$

$$C_5H_5Co(CO)_2 + R_fI \xrightarrow{50°} C_5H_5Co(CO)R_fI + CO$$

$$[C_5H_5NiCO]_2 + R_fI \longrightarrow R_fNi(CO)C_5H_5 + C_5H_5Ni(CO)I$$
$$\text{(decomposes further)}$$

Polyfluoroalkyl transition-metal derivatives with metal–carbon σ-bonds may be prepared by addition of a transition-metal hydride to a fluoroolefin, e.g.

$$HMn(CO)_5 + CF_2{=}CF_2 \longrightarrow HCF_2CF_2Mn(CO)_5$$

TRANSITION-METAL COMPLEXES OF CARBORANE CAGES [58]

The development of polyhedral carborane chemistry in the last few years has led to the discovery of polyhedral systems with a relatively large single pentagonal face of similar dimensions to the π-C_5H_5 (cyclopentadienyl) ring. These polyhedral carboranes can form a bond with appropriate transition metals similar to the metal–π-cyclopentadienyl bond. Carborane cage complexes are now known for the carboranes $B_6C_2H_8^{2-}$, $B_7C_2H_9^{2-}$, $B_9C_2H_{11}^{2-}$, $B_{10}CH_{11}^{3-}$, and $B_{10}H_{10}CNR_3^{2-}$. The relatively high negative charges of 2– and 3– on these carborane cages known to complex with transition metals mean that these cages stabilize higher formal oxidation states of the metal atom than does the cyclopentadienide anion with a charge of 1–.

Among the various carborane polyhedra cited above, the greatest number of metal complexes have been made from the eleven-particle icosahedral fragment $B_9C_2H_{11}^{2-}$ depicted schematically in structure **IV**. The transition

IV

metal provides the twelfth atom necessary to close the icosahedron. The following equations illustrate some of the reactions used to prepare $B_9C_2H_{11}^{2-}$ complexes of transition metals.

a. *Preparation of ligand*

$$B_{10}H_{14} + HC\equiv CH \xrightarrow{\text{acetonitrile}} 1,2\text{-}B_{10}C_2H_{12} + 2\ H_2\uparrow$$

$$B_{10}C_2H_{12} + OH^- + 2\ H_2O \longrightarrow B_9C_2H_{12}^- + H_3BO_3 + H_2\uparrow$$

$$B_9C_2H_{12}^- \underset{+H^+}{\overset{-H^+}{\rightleftharpoons}} B_9C_2H_{11}^{2-}$$

b. *Preparation of metal derivatives*

$$2\ B_9C_2H_{11}^{2-} + MCl_2 \longrightarrow [(B_9C_2H_{11})_2M]^{2-} + 2\ Cl^-$$
$$(M = Cr,\ Fe,\ Co,\ Ni,\ Cu)$$

$$3\ [(B_9C_2H_{11})_2Co]^{2-} \xrightarrow{\text{spontaneous}} 2\ [(B_9C_2H_{11})_2Co]^- + 2\ B_9C_2H_{11}^{2-} + Co$$

$$[(B_9C_2H_{11})_2M]^{2-} \xrightarrow{\text{air}} [(B_9C_2H_{11})_2M]^- + e^-$$
$$(M = Fe,\ Ni)$$

$$[(B_9C_2H_{11})_2Ni]^- + Fe^{3+} \longrightarrow (B_9C_2H_{11})_2Ni + Fe^{2+}$$
$$\text{(yellow-orange,}$$
$$\text{volatile)}$$

$$B_9C_2H_{11}^{2-} + M(CO)_5Br \longrightarrow [(B_9C_2H_{11})M(CO)_3]^- + Br^-$$
$$(M = Mn\ or\ Re)$$

The complex anions can be isolated as cesium or tetramethylammonium salts.

Summary of Chapter I and Further Plan of This Volume

This introductory chapter on transition-metal organometallic chemistry has provided a general outline of the principles important to this area of organo-metallic chemistry. Furthermore, some of the important bonding principles and preparative techniques in the study and application of transition-metal organometallic compounds have been discussed. Areas receiving particular attention have been the metal carbonyls; metal derivatives of various C_nH_n ligands, including π-cyclopentadienyl; metal complexes of olefins, poly-olefins, and acetylenes; metal π-allyl derivatives; compounds with transition-metal–carbon σ-bonds; and transition-metal complexes of polyhedral carborane cages.

In the remaining chapters of this volume descriptive transition-metal organometallic chemistry organized according to the metal covered and its position in the periodic table will be discussed. This descriptive chemistry will provide further illustrations of the general principles outlined in this first chapter.

REFERENCES

1. W. C. Zeise, *Pogg. Annalen* [2] **9**, 632 (1827).
2. L. Mond, C. Langer, and F. Quincke, *J. Chem. Soc.* **57**, 749 (1890).
3. F. Hein, *Ber.* **52**, 195 (1919); F. A. Cotton, *Chem. Rev.* **55**, 551 (1955).
4. E. O. Fischer and W. Hafner, *Z. Naturforsch.* **10b**, 665 (1955).
5. T. J. Kealy and P. L. Pauson, *Nature* **168**, 1039 (1951).
6. S. A. Miller, J. A. Tebboth, and J. F. Tremaine, *J. Chem. Soc.* p. 632 (1952).
7. R. B. King, *Adv. Chem. Ser.* **62**, 203 (1967).
8. E. H. Braye, L. F. Dahl, W. Hübel, and D. L. Wampler, *J. Am. Chem. Soc.* **84**, 4633 (1962).
9. B. F. G. Johnson, R. J. Johnston, and J. Lewis, *Chem. Commun.* p. 1057 (1967).
10. W. R. Gilliland and A. A. Blanchard, *Inorg. Syn.* **2**, 234 (1946).
11. W. Hieber and H. Fuchs, *Z. Anorg. Allgem. Chem.* **248**, 256 (1941); C. W. Bradford and R. S. Nyholm, *Chem. Commun.* p. 384 (1967).
12. I. Wender, H. W. Sternberg, S. Metlin, and M. Orchin, *Inorg. Syn.* **5**, 190 (1957).
13a. R. P. M. Werner and H. E. Podall, *Chem. & Ind.* (*London*) p. 144 (1961); b R. P. M. Werner, A. H. Filbey, and S. A. Manastyrskyj, *Inorg. Chem.* **3**, 298 (1964); c H. E. Podall, H. B. Prestridge, and H. Shapiro, *J. Am. Chem. Soc.* **83**, 2057 (1961).
14. G. Natta, R. Ercoli, F. Calderazzo, and A. Rabizzoni, *J. Am. Chem. Soc.* **79**, 3611 (1957); R. Ercoli, F. Calderazzo, and G. Bernardi, *Gazz. Chim. Ital.* **89**, 809 (1959).
15. E. O. Fischer, W. Hafner, and K. Öfele, *Ber.* **92**, 3050 (1959).
16. K. A. Kocheskov, A. N. Nesmeyanov, M. M. Nadj, and I. M. Rossinskaya, *Compt. Rend. Acad. Sci. URSS* **26**, 54 (1940).
17. E. R. Corey, L. F. Dahl, and W. Beck, *J. Am. Chem. Soc.* **85**, 1202 (1963); W. Hieber, R. Schuh, and H. Fuchs, *Z. Anorg. Allgem. Chem.* **248**, 243 (1941).
18. A. Job and A. Cassal, *Compt. Rend.* **183**, 392 (1926); A. Job and J. Rouvillois, *Bull. Soc. Chim. France* **41**, 1041 (1927); W. Hieber and E. Romberg, *Z. Anorg. Allgem. Chem.* **221**, 321 (1935); B. B. Owen, J. English, Jr., H. G. Cassidy, and C. V. Dundon, *J. Am. Chem. Soc.* **69**, 1723 (1947).
19. H. E. Podall, J. H. Dunn, and H. Shapiro, *J. Am. Chem. Soc.* **82**, 1325 (1960).
20. A. N. Nesmeyanov, K. N. Anisimov, V. L. Volkov, A. E. Fridenberg, E. P. Mihkeev, and A. V. Medvedeva, *Zh. Neorgan. Khim.* **4**, 1827 (1959); *Chem. Abstr.* **54**, 11795g (1960).
21. R. D. Closson, L. R. Buzbee, and G. C. Ecke, *J. Am. Chem. Soc.* **80**, 6167 (1958).
22. A. N. Nesmeyanov, E. P. Mihkeev, K. N. Anisimov, V. L. Volkov, and Z. P. Valueva, *Zh. Neorgan. Khim.* **4**, 503, 249 (1959); *Chem. Abstr.* **53**, 21327h and 12907c (1959).
23. G. Wilkinson, F. A. Cotton, and J. M. Birmingham, *J. Inorg. & Nucl. Chem.* **2**, 95 (1956).
24. W. A. Barber, *Inorg. Syn.* **6**, 11 (1960); C. L. Sloan and W. A. Barber, *J. Am. Chem. Soc.* **81**, 1364 (1959); A. F. Reid and P. C. Wailes, *J. Organometal. Chem.* (*Amsterdam*) **2**, 329 (1964); A. F. Reid and P. C. Wailes, *Australian J. Chem.* **19**, 309 (1966).
25. A. N. Nesmeyanov, R. B. Materikova, and N. S. Kochetkova, *Izv. Akad. Nauk SSSR, Ser. Khim.* p. 1334 (1963); *Chem. Abstr.* **59**, 12841d (1963); C. C. Hunt and J. R. Doyle, *Inorg. Nucl. Chem. Letters* **2**, 283 (1966); T. J. Katz and J. J. Mrowca, *J. Am. Chem. Soc.* **89**, 1105 (1967); R. B. King, *Inorg. Chem.* **7**, 90 (1968).
26. G. Wilkinson, *Org. Syn.* **36**, 31 (1956); R. B. King, *Organometal. Syn.* **1**, 71 (1965).
27. R. B. King and A. Fronzaglia, *Inorg. Chem.* **5**, 1837 (1966).

28. T. S. Piper, F. A. Cotton, and G. Wilkinson, *J. Inorg. & Nucl. Chem.* **1**, 165 (1955); R. B. King, *Organometal. Syn.* **1**, 114–119 (1965).
29. T. S. Piper and G. Wilkinson, *J. Inorg. & Nucl. Chem.* **3**, 104 (1956).
30. H. Breil and G. Wilke, *Angew. Chem. Intern. Ed. Engl.* **5**, 898 (1966); H. Dietrich and H. Dierks, *ibid.* p. 899.
31. E. W. Gowling and S. F. A. Kettle, *Inorg. Chem.* **3**, 604 (1964).
32. R. Criegee and G. Schröder, *Ann.* **623**, 1 (1959).
33. G. F. Emerson, L. Watts, and R. Pettit, *J. Am. Chem. Soc.* **87**, 131 (1965); R. G. Amiet, P. C. Reeves, and R. Pettit, *Chem. Commun.* p. 1208 (1967).
34. M. Rosenblum and C. Gatsonis, *J. Am. Chem. Soc.* **89**, 5074 (1967); M. Rosenblum and B. North, *ibid.* **90**, 1060 (1968).
35. R. Bruce, K. Moseley, and P. M. Maitlis, *Can. J. Chem.* **45**, 2011 (1967); R. Bruce and P. M. Maitlis, *Can. J. Chem.* **45**, 2018 (1967).
36. P. M. Maitlis and M. L. Games, *J. Am. Chem. Soc.* **85**, 1887 (1963); *Chem. & Ind.* (*London*) p. 1624 (1963); P. M. Maitlis, A. Efraty, and M. L. Games, *J. Organometal. Chem.* (*Amsterdam*) **2**, 284 (1964).
37. E. O. Fischer, K. Öfele, H. Essler, W. Fröhlich, J. P. Mortensen, and W. Semmlinger, *Z. Naturforsch.* **13b**, 458 (1958); *Ber.* **91**, 2763 (1958); B. Nicholls and M. C. Whiting, *J. Chem. Soc.* p. 551 (1959); G. Natta, R. Ercoli, F. Calderazzo, and S. Santambrogio, *Chim. Ind.* (*Milan*) **40**, 1003 (1958).
38. F. Calderazzo, *Inorg. Chem.* **3**, 1207 (1964).
39. G. Winkhaus, L. Pratt, and G. Wilkinson, *J. Chem. Soc.* p. 3807 (1961); E. O. Fischer and F. J. Kohl, *Z. Naturforsch.* **18b**, 504 (1963).
40. E. O. Fischer and H. P. Kögler, *Ber.* **90**, 250 (1957).
41. M. Tsutsui and H. H. Zeiss, *J. Am. Chem. Soc.* **81**, 1367 (1959).
42. E. O. Fischer and S. Breitschaft, *Ber.* **99**, 2213 (1966).
43. W. Hübel and C. Hoogzand, *Ber.* **93**, 103 (1960).
44. E. O. Fischer and J. Müller, *Z. Naturforsch.* **17b**, 776 (1962).
45. R. B. King and F. G. A. Stone, *J. Am. Chem. Soc.* **81**, 5263 (1959).
46. R. P. M. Werner and S. A. Manastyrskyj, *J. Am. Chem. Soc.* **83**, 2023 (1961); F. Calderazzo and P. L. Calvi, *Chim. Ind.* (*Milan*) **44**, 1217 (1962).
47. H. J. Dauben, Jr. and L. R. Honnen, *J. Am. Chem. Soc.* **80**, 5570 (1958).
48. D. P. Tate, J. M. Augl, W. M. Ritchey, B. L. Ross, and J. G. Grasselli, *J. Am. Chem. Soc.* **86**, 3261 (1964); R. Tsumura and N. Hagihara, *Bull. Chem. Soc. Japan* **38**, 1901 (1965).
49. H. Greenfield, H. W. Sternberg, R. A. Friedel, J. H. Wotiz, R. Markby, and I. Wender, *J. Am. Chem. Soc.* **78**, 120 (1956); J. F. Tilney-Bassett, *J. Chem. Soc.* p. 577 (1961).
50. G. Wilke, B. Bogdanović, P. Hardt, P. Heimbach, W. Keim, M. Kröner, W. Oberkirch, K. Tanaka, E. Steinrücke, D. Walter, and H. Zimmermann, *Angew. Chem. Intern. Ed. Engl.* **5**, 151 (1966).
51. R. A. Plowman and F. G. A. Stone, *Z. Naturforsch.* **17b**, 575 (1962); R. F. Heck and D. S. Breslow, *J. Am. Chem. Soc.* **82**, 750 (1960); J. Powell and B. L. Shaw, *J. Chem. Soc., A* p. 583 (1968).
52. H. B. Jonassen, R. I. Stearns, J. Kenttämaa, D. W. Moore, and A. G. Whittaker, *J. Am. Chem. Soc.* **80**, 2586 (1958); W. R. McClellan, H. H. Hoehn, H. N. Cripps, E. L. Muetterties, and B. W. Howk, *ibid.* **83**, 1601 (1961).
53. R. Hüttel, H. Dietl, and H. Christ, *Ber.* **97**, 2037 (1964).
54. G. F. Emerson, K. Ehrlich, W. P. Giering, and P. C. Lauterbur, *J. Am. Chem. Soc.* **88**, 3172 (1966).
55. A. Nakamura and N. Hagihara, *J. Organometal. Chem.* (*Amsterdam*) **3**, 480 (1965).

56. F. A. Cotton, *Chem. Rev.* **55**, 551 (1955).
57. P. M. Treichel and F. G. A. Stone, *Advan. Organometal. Chem.* **1**, 143 (1964).
58. M. F. Hawthorne, D. C. Young, T. D. Andrews, T. V. Howe, R. L. Pilling, A. D. Pitts, M. Reintjes, L. F. Warren, Jr., and P. A. Wegner, *J. Am. Chem. Soc.* **90**, 879 (1968).

SUPPLEMENTARY READING

1. E. W. Abel, The metal carbonyls. *Quart. Rev.* **17**, 133 (1963).
2. J. C. Hileman, Metal carbonyls. *Preparative Inorg. Reactions* **1**, 77 (1964).
3. R. B. King, Reactions of alkali-metal derivatives of metal carbonyls and related compounds. *Advan. Organometal. Chem.* **2**, 157 (1964).
4. T. A. Manuel, Lewis base–metal carbonyl complexes. *Advan. Organometal. Chem.* **3**, 181 (1965).
5. W. Strohmeier, Photochemical substitutions on metal carbonyls and their derivatives. *Angew. Chem. Intern. Ed. Engl.* **3**, 730 (1964).
6. R. L. Pruett, Cyclopentadienyl and arene metal carbonyls. *Preparative Inorg. Reactions* **2**, 187 (1965).
7. G. Wilkinson and F. A. Cotton, Cyclopentadienyl and arene metal compounds. *Progr. Inorg. Chem.* **1**, 1 (1959).
8. E. O. Fischer and H. P. Fritz, Complexes of aromatic ring systems and metals. *Advan. Inorg. Chem. Radiochem.* **1**, 55 (1959).
9. K. Plesske, Ring substitutions and secondary reactions of aromatic–metal π-complexes. *Angew. Chem. Intern. Ed. Engl.* **1**, 312 and 394 (1962).
10. J. Birmingham, Synthesis of cyclopentadienyl metal compounds. *Advan. Organometal. Chem.* **2**, 366 (1964).
11. P. M. Maitlis, Cyclobutadiene–metal complexes. *Advan. Organometal. Chem.* **4**, 95 (1966).
12. M. A. Bennett, Metal π-complexes formed by seven-membered and eight-membered carbocyclic compounds. *Advan. Organometal. Chem.* **4**, 353 (1966).
13. E. O. Fischer and H. Werner, Metal π-complexes with di- and oligo-olefinic ligands. *Angew. Chem. Intern. Ed. Engl.* **2**, 80 (1963).
14. M. A. Bennett, Olefin and acetylene complexes of transition metals. *Chem. Rev.* **62**, 611 (1962).
15. R. G. Guy and B. L. Shaw, Olefin, acetylene, and π-allylic complexes of transition metals. *Advan. Inorg. Chem. Radiochem.* **4**, 78 (1962).
16. M. L. H. Green and P. L. I. Nagy, Allyl metal complexes. *Advan. Organometal. Chem.* **2**, 330 (1964).
17. G. Wilke, B. Bogdanović, P. Hardt, P. Heimbach, W. Keim, M. Kröner, W. Oberkirch, K. Tanaka, E. Steinrücke, D. Walter, and H. Zimmermann, Allyl transition metal systems. *Angew. Chem. Intern. Ed. Engl.* **5**, 151 (1966).
18. P. M. Treichel and F. G. A. Stone, Fluorocarbon derivatives of metals. *Advan. Organometal. Chem.* **1**, 143 (1964).
19. M. F. Hawthorne, The chemistry of the polyhedral species derived from transition metals and carboranes. *Accounts Chem. Res.* **1**, 257 (1968).

QUESTIONS

1. The neutral nitrosyl ligand (NO) behaves as a three-electron donor. On the basis of electronic configuration and coordination number predict the formulas of the first-row ($3d$) transition-metal derivatives containing only carbonyl and nitrosyl groups.

2. Suggest possible methods for the preparation of stable arylazo derivatives of transition metals.

3. Phosphorus trifluoride (PF_3) is a ligand which has "back-bonding" abilities similar to carbon monoxide, but which forms stronger chemical bonds to transition metals than carbon monoxide. Predict the formulas of the stable phosphorus trifluoride (trifluorophosphine) derivatives of transition metals. Suggest possible preparative methods for metal trifluorophosphine complexes, keeping in mind the gaseous nature of phosphorus trifluoride (bp $-101°$) and its possible sensitivity to strong reducing agents such as the electropositive metals and their organometallic derivatives.

4. Summarize the bonding of nitric oxide to a metal atom by a table similar to Table I-1 for the bonding of carbon monoxide to a metal atom.

5. Give examples of complexes of the indenyl, azulene, and pyrrolyl ligands, where the five-membered ring is bonded to the metal atom in a manner similar to the π-cyclopentadienyl ligand–metal complexes. Suggest preparative methods for the complexes suggested as examples.

6. Discuss the fact that a cyclopentadienylpalladium carbonyl has not yet been reported (see Table I-9).

7. Suggest preparative methods for five of the cyclopentadienylmetal carbonyls of Table I-9 which are not discussed in this section of the text.

8. Discuss the bonding of benzene to a metal atom using drawings analogous to those in Fig. I-2 for the π-cyclopentadienylmetal bond.

9. Suggest preparative methods for bis(cyclobutadiene)nickel, a presently (1968) unknown complex.

10. Account for the inability to prepare $Cr(CO)_3$ complexes of hexafluorobenzene and of nitrobenzene.

11. Give examples of systems where nonconjugated diolefins form different types of complexes than conjugated diolefins.

12. Suggest preparations of compounds of the types listed in Table I-11.

13. Discuss possible preparative methods for σ-vinyl transition-metal derivatives.

14. Suggest possible new metal derivatives of the trimethylenemethane and bisallylene ligands. Suggest possible ways for preparing these compounds.

CHAPTER II

Organometallic Derivatives of the Early Transition Metals

Introduction

The early transition metals are most often considered to be those which require more than twelve electrons (i.e., six two-electron donor ligands) to attain the electronic configuration of the next rare gas (see Table I-1, Chapter I). Thus, the early transition metals consist of scandium, yttrium, the lanthanides, the actinides, titanium, zirconium, hafnium, vanadium, niobium, and tantalum. Since the organometallic chemistry of the early transition metals is relatively limited compared with that of the other transition metals, it is convenient to discuss the organometallic chemistry of all of the early transition metals in one chapter.

Organometallic Chemistry of the Lanthanides (Including Scandium and Yttrium)

The organometallic chemistry of these metals is limited to the cyclopentadienyls, which appear to be ionic derivatives containing metal cations and cyclopentadienide anions. The lanthanides are normally trivalent and, therefore, form cyclopentadienides $(C_5H_5)_3M$. These derivatives are prepared by reaction of the anhydrous lanthanide trichloride with an alkali-metal cyclopentadienide (NaC_5H_5 in tetrahydrofuran or KC_5H_5 in benzene) [1–3]. The lanthanide tricyclopentadienides, $(C_5H_5)_3M$, are air-sensitive solids, sublimable at 200°–250° in high vacuum, and exhibit the characteristic colours and magnetisms of the tripositive lanthanide ions. They hydrolyze rapidly in water to give free cyclopentadiene and the lanthanide oxide or hydroxide. The lanthanide tricyclopentadienides react with various Lewis bases, such as ammonia, tetrahydrofuran, cyclohexyl isocyanide, and triphenylphosphine, to give adducts of the general formula $(C_5H_5)_3M \leftarrow L$. The derivatives of the lower molecular weight Lewis bases including cyclohexyl isocyanide sublime in high vacuum. A few examples of trivalent lanthanide cyclopentadienide chlorides of the types $C_5H_5MCl_2$ and $(C_5H_5)_2MCl$ are known.

Europium and ytterbium form not only the usual tripositive cations but also dipositive cations. Their divalent cyclopentadienides can be prepared by reaction of the metal with cyclopentadiene in liquid ammonia according to the following equation [4]:

$$M + 2 C_5H_6 \longrightarrow (C_5H_5)_2M + H_2$$

where $M = Eu$, yellow, 7.63 BM; $M = Yb$, red, diamagnetic.

Organometallic Chemistry of the Actinides

The organometallic chemistry of the actinides, like that of the lanthanides, was limited until very recently to the cyclopentadienyl derivatives, which are believed to be ionic cyclopentadienides. The known cyclopentadienide derivatives of thorium, uranium, and neptunium have the metal in the 4+ oxidation state, whereas the known cyclopentadienide derivatives of plutonium and americium have the metal in the 3+ oxidation state. The preparation and properties of the actinide cyclopentadienides are summarized in the following equations.

THORIUM DERIVATIVES

$$4 C_5H_5Na + ThCl_4 \longrightarrow (C_5H_5)_4Th + 4 NaCl \quad [5] \tag{a}$$
(white, sublimes
190°/high vacuum)

$$3 C_5H_5K + ThCl_4 \xrightarrow{ether} (C_5H_5)_3ThCl + 3 KCl \quad [6] \tag{b}$$
(white, sublimes
200°/10^{-4} mm)

$$3 C_5H_5Na + ThCl_4 + NaOR \longrightarrow (C_5H_5)_3ThOR + 4 NaCl \quad [6] \tag{c}$$
$(R = CH_3$ or $C_4H_9)$ (white)

All thorium cyclopentadienide compounds are diamagnetic and hydrolyzed by water.

URANIUM DERIVATIVES

$$4 C_5H_5Na + UCl_4 \longrightarrow (C_5H_5)_4U + 4 NaCl \quad [7] \tag{a}$$
(red,
nonvolatile)

$$3 C_5H_5Na + UCl_4 + NaOR \longrightarrow (C_5H_5)_3UOR + 4 NaCl \quad [8] \tag{b}$$
(green)

$$(R = CH_3 \text{ or } C_4H_9)$$

DERIVATIVES OF HEAVIER ACTINIDES

$$2 \ NpCl_4 + 3 \ Be(C_5H_5)_2 \xrightarrow{70°} 2 \ (C_5H_5)_3NpCl + 3 \ BeCl_2 \qquad [9a]$$
(dark brown)

$$2 \ PuCl_3 + 3 \ Be(C_5H_5)_2 \xrightarrow{70°} 2 \ (C_5H_5)_3Pu + 3 \ BeCl_2 \qquad [9b]$$
(moss green)

$$2 \ AmCl_3 + 3 \ Be(C_5H_5)_2 \xrightarrow{70°} 2 \ (C_5H_5)_3Am + 3 \ BeCl_2 \qquad [9c]$$
(flesh, glows
in the dark)

These reactions illustrate the use of molten beryllium cyclopentadienide for the introduction of cyclopentadienyl groups.

A novel type of organouranium compound is bis(cyclooctatetraene)uranium, $(C_8H_8)_2U$, which may be prepared by the following reaction [10a]:

$$UCl_4 + 2 \ K_2C_8H_8 \xrightarrow{THF} (C_8H_8)_2U + 4 \ KCl$$

This compound is a pyrophoric green solid subliming at 180°/0.03 mm and stable to water, acetic acid, and aqueous sodium hydroxide. A novel "sandwich" structure (**A**) has been postulated for $(C_8H_8)_2U$. The metal–ring bonding in $(C_8H_8)_2U$ probably involves the uranium f orbitals in addition to the more usual s, p, and d orbitals.

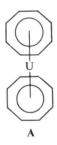

A

Organometallic Chemistry of Titanium—π-Cyclopentadienyl Derivatives

Most of the organometallic chemistry of titanium is concerned with derivatives containing π-cyclopentadienyl ligands. In addition, a few unusual cyclooctatetraene derivatives and some unstable but catalytically important alkyltitanium derivatives with titanium–carbon σ-bonds are known.

The most important of the cyclopentadienyltitanium derivatives is biscyclopentadienyltitanium dichloride, $(C_5H_5)_2TiCl_2$ (**I**: M = Ti, X = Cl), readily obtained by treatment of titanium tetrachloride with a stoichiometric quantity of sodium cyclopentadienide in tetrahydrofuran solution [*10b*]. This compound forms red, air-stable crystals, mp 289°–291°, which are insoluble in water but soluble to varying extents in polar organic solvents. This titanium compound, $(C_5H_5)_2TiCl_2$, is an extremely useful intermediate for the preparation of other organotitanium derivatives. Interesting organotitanium compounds which can be prepared from $(C_5H_5)_2TiCl_2$ include derivatives of the following types.

COMPOUNDS WITH TITANIUM–CARBON σ-BONDS [*11*]

$$(C_5H_5)_2TiCl_2 + 2\ LiR \xrightarrow{\text{ethers}} (C_5H_5)_2TiR_2 + 2\ LiCl$$
$$\text{(orange)}$$

where $R = C_6H_5$ or CH_3.

A CARBONYL DERIVATIVE OF TITANIUM [*12*]

The only known compound with titanium–carbonyl bonds is red-brown, air-sensitive $(C_5H_5)_2Ti(CO)_2$ (I: M = Ti, X = CO) which can be obtained by reaction of $(C_5H_5)_2TiCl_2$ with carbon monoxide at elevated temperatures in the presence of a strong reducing agent, such as butyllithium or sodium cyclopentadienide. The carbonyl groups in $(C_5H_5)_2Ti(CO)_2$ exhibit strong $\nu(CO)$ frequencies at 1968 and 1889 cm^{-1}, the normal region for terminal carbonyl groups attached directly to a metal atom.

π-ALLYL DERIVATIVES OF TITANIUM [*13*]

Purple, air-sensitive titanium derivatives of the type $(C_5H_5)_2Ti(allyl)$ (**II**: R = H or CH_3) have been prepared either by reaction of allylmagnesium chloride with $(C_5H_5)_2TiCl_2$ (**I**: X = Cl) in diethyl ether solution or by ultraviolet irradiation of a diethyl ether solution containing $(C_5H_5)_2TiCl_2$, isopropylmagnesium bromide (as a reducing agent), and butadiene (**II**: R = CH_3).

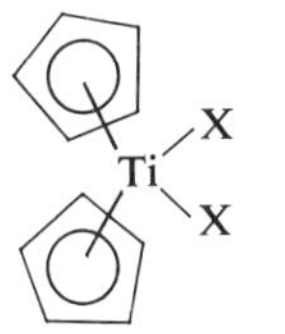

I 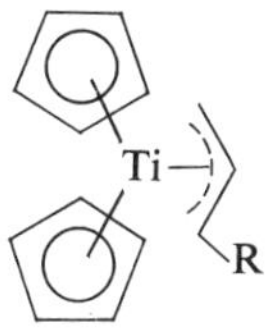   II

CYCLOPENTADIENYLTITANIUM TRICHLORIDE

A redistribution (or symproportionation) reaction between $(C_5H_5)_2TiCl_2$ and titanium tetrachloride produces yellow, readily hydrolyzed cyclopentadienyltitanium trichloride, $C_5H_5TiCl_3$ (**III**: Y = Cl) mp 208°–211°, sublimes 100°/0.1 mm, according to the following equation [14]:

$$(C_5H_5)_2TiCl_2 + TiCl_4 \quad \xrightarrow[140°, 2\frac{1}{4}\ hr]{xylene} \quad 2\ C_5H_5TiCl_3$$

III

The following sequence of reactions can also be used to prepare $C_5H_5TiCl_3$ [15]:

$$3\ Ti(OR)_4 + TiCl_4 \quad \xrightarrow{THF} \quad 4\ ClTi(OR)_3 \qquad (a)$$

(R = ethyl or isopropyl)

$$ClTi(OR)_3 + C_5H_5Na \quad \xrightarrow{THF} \quad C_5H_5Ti(OR)_3 \qquad (b)$$
(volatile colorless liquid)

$$C_5H_5Ti(OR)_3 + 3\ CH_3COCl \quad \xrightarrow{CHCl_3} \quad C_5H_5TiCl_3 + 3\ CH_3COOR \qquad (c)$$

Reactions of $C_5H_5TiCl_3$ with alkyllithium compounds in diethyl ether solution at low temperatures give fairly unstable alkyltitanium derivatives of the type $C_5H_5TiR_3$ (**III**: Y = CH_3, yellow solid [16]; Y = C_6H_5, orange plates [17]) which contain three titanium–carbon σ-bonds.

TITANIUM(III) CYCLOPENTADIENYL DERIVATIVES

Reactions of $(C_5H_5)_2TiCl_2$ with various strong reducing agents can be used to prepare π-cyclopentadienyl derivatives of titanium(III) with one, two, or three cyclopentadienyl groups per titanium atom. All of these compounds are extremely sensitive to air oxidation. The following reactions exemplify some of the preparative techniques:

$$(C_5H_5)_2TiCl_2 + 2\ NaC_5H_5 \quad \xrightarrow{THF} \quad (C_5H_5)_3Ti + NaCl + \text{other products} \quad [18] \quad (a)$$
(green, sublimes 125°/high vacuum, 1.69 BM)

$$2 \; (C_5H_5)_2TiCl_2 + Zn \quad \longrightarrow \quad [(C_5H_5)_2TiCl]_2 + ZnCl_2 \quad [19] \qquad (b)$$
(green to violet-brown, mp 282–283°,
sublimes 170°/high vacuum, 1.56 BM)

$$(C_5H_5)_2TiCl_2 + Al(C_2H_5)_3 \quad \longrightarrow \quad C_5H_5TiCl_2 + \text{other products} \qquad [20] \qquad (c)$$
(blue)

The titanium compound in (b) appears to have the binuclear structure **IV** with chlorine bridges. It can also be prepared by heating titanium trichloride with magnesium cyclopentadienide [19].

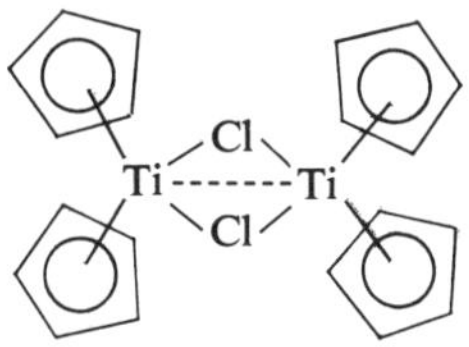

IV

These titanium(III) derivatives are paramagnetic with values close to the 1.73 BM expected for one unpaired electron. Deviations from this expected value in some of the binuclear derivatives are attributed to titanium–titanium bonding.

BISCYCLOPENTADIENYLTITANIUM(II)

In 1955 the reaction between titanium dichloride and sodium cyclopentadienide was reported to give green, volatile, diamagnetic, extremely airsensitive biscyclopentadienyltitanium(II), $(C_5H_5)_2Ti$, believed at that time to be an analogue of ferrocene [21]. This reaction proved very difficult to reproduce in other laboratories. However, in 1966 Watt, Baye, and Drummond [22] developed a more reliable synthesis of biscyclopentadienyltitanium(II) from $(C_5H_5)_2TiCl_2$ and sodium naphthalenide as a reducing agent:

$$(C_5H_5)_2TiCl_2 + 2 \; NaC_{10}H_8 \quad \xrightarrow{\text{THF}} \quad (C_5H_5)_2Ti + 2 \; C_{10}H_8 + 2 \; NaCl$$

These workers found biscyclopentadienyltitanium(II) to be a green diamagnetic, extremely air-sensitive solid, volatile at 140°/0.2 mm and dimeric in solution. The diamagnetism in biscyclopentadienyltitanium(II) may arise from spin-pairing through titanium–titanium bonding, rather than from a fundamental property of the monomeric $(C_5H_5)_2Ti$ unit.

NITROGEN-FIXING π-CYCLOPENTADIENYLTITANIUM DERIVATIVES

The process of fixing elemental nitrogen is of importance both to biological systems and for the economical synthesis of important nitrogen compounds.

In 1966 Vol'pin and Shur [23] reported that a mixture of $(C_5H_5)_2TiCl_2$ and ethylmagnesium bromide reacted with elemental nitrogen at 90°–100°/150 atm to give a mixture which gave ammonium ion upon acid hydrolysis. On the basis of ESR studies, Brintzinger [24] suggested that the nitrogen-fixing species in this $(C_5H_5)_2TiCl_2/C_2H_5MgBr$ was the titanium(III) hydride derivative $[(C_5H_5)_2TiH]_2$ of proposed structure V. Other workers [24a] find titanium(II) derivatives, such as biscyclopentadienyltitanium(II), to be active nitrogen-fixing species.

V

Other Organometallic Derivatives of Titanium

Mixtures of titanium halides and aluminum alkyls are of importance as Ziegler–Natta catalysts for the polymerization of ethylene and related olefins. The active ingredient in these catalysts appears to be an unstable alkyltitanium derivative with a titanium–carbon σ-bond. The importance of these polymerization catalysts makes the chemistry of alkyltitanium derivatives of particular interest. The instability of many compounds with titanium–carbon σ-bonds limits the range of alkyltitanium derivatives which can be studied. Increasing the number of alkyl groups bonded to the titanium atom appears to lower the stability of the compound. All of the alkyltitanium compounds are sensitive to air and water.

The methyltitanium compounds illustrate some types of compounds with titanium–carbon σ-bonds. Treatment of titanium tetrachloride with tetramethyllead replaces one chlorine atom with a methyl group, giving violet methyltitanium trichloride, CH_3TiCl_3, mp 28–29°, dec. >30° [25]. Treatment of titanium tetrachloride with methyllithium, a much stronger alkylating agent, replaces all four chlorine atoms with methyl groups, giving bright-yellow tetramethyltitanium, $(CH_3)_4Ti$ [26]. Tetramethyltitanium is very unstable and decomposes when warmed above −78°. It can be stabilized by complexing with a chelating diamine (o-phenanthroline or 2,2'-bipyridyl) to form the orange to red octahedral $(CH_3)_4Ti$(amine) complexes decomposing only above 0° [27].

Phenyltitanium compounds can also be prepared [28]. Reaction between titanium tetrachloride and phenyllithium in diethyl ether solution at −70° gives orange, unstable tetraphenyltitanium, $(C_6H_5)_4Ti$. When warmed to room temperature, tetraphenyltitanium decomposes with the elimination of biphenyl, according to the following equation:

$$(C_6H_5)_4Ti \xrightarrow{25°} \frac{1}{m} [(C_6H_5)_2Ti]_m + C_6H_5\text{—}C_6H_5$$

The diphenyltitanium(II) decomposition product is a black pyrophoric solid which has not been studied in much detail but which does appear to be polymeric.

Two unusual cyclooctatetraene derivatives of titanium have been prepared from tetrabutyltitanate, cyclooctatetraene, and excess triethylaluminum [29]. If the molar ratio of cyclooctatetraene to $Ti(OC_4H_9)_4$ is 2:1, yellow $Ti_2(C_8H_8)_3$ is obtained. This complex is shown by X-ray diffraction to have structure **VI** with two of the three C_8H_8 rings in the unusual planar configuration [30]. If the molar ratio of cyclooctatetraene to $Ti(OC_4H_9)_4$ is raised to 10:1, then red $Ti(C_8H_8)_2$ of unknown structure is obtained. Both of these cyclooctatetraene titanium derivatives are very sensitive to water and oxygen, and even react with carbon dioxide under pressure with loss of the titanium to form carboxylic acids containing eight-membered carbocyclic rings.

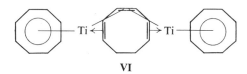

VI

Organometallic Chemistry of Zirconium and Hafnium

Most of the organometallic chemistry of zirconium and hafnium is concerned with π-cyclopentadienyl derivatives. Reactions of the metal tetrahalides with sodium cyclopentadienide give first $(C_5H_5)_2MCl_2$ and then $(C_5H_5)_4M$, according to the following equations (M = Zr or Hf):

$$MCl_4 + 2\ NaC_5H_5 \longrightarrow (C_5H_5)_2MCl_2 + 2\ NaCl \qquad [10b] \qquad \text{(a)}$$
$$\text{(white)}$$

$$(C_5H_5)_2MCl_2 + 2\ NaC_5H_5 \longrightarrow (C_5H_5)_4M + 2\ NaCl \qquad [31] \qquad \text{(b)}$$
$$\text{(white)}$$

The white colors of the zirconium and hafnium compounds $(C_5H_5)_2MCl_2$ (M = Zr or Hf) contrast with the red color of the titanium analogue $(C_5H_5)_2TiCl_2$, exemplifying the generally observed trend of lighter colors of analogous transition-metal organometallic compounds when descending a column of the periodic table.

The zirconium compound $(C_5H_5)_2ZrCl_2$ has been used as a precursor for the preparation of other π-cyclopentadienylzirconium compounds. Thus the borohydride derivative $(C_5H_5)_2Zr(BH_4)_2$ can be prepared by reaction with lithium borohydride, according to the following equation [32]:

$$(C_5H_5)_2ZrCl_2 \ + \ 2 \ LiBH_4 \ \xrightarrow{\text{ether}} \ (C_5H_5)_2Zr(BH_4)_2 \ + \ 2 \ LiCl$$

(white, diamagnetic,
volatile at 110°/0.1 mm)

This zirconium borohydride derivative has the metal in the 4+ oxidation state; the reaction between $(C_5H_5)_2ZrCl_2$ and lithium borohydride does not lead to reduction of the metal atom. By contrast the titanium derivative $(C_5H_5)_2TiCl_2$ is reduced by lithium borohydride to a titanium borohydride derivative $(C_5H_5)_2TiBH_4$ with the metal in the 3+ oxidation state, according to the following equation [33]:

$$2 \ (C_5H_5)_2TiCl_2 \ + \ 4 \ LiBH_4 \ \xrightarrow{\text{ether}} \ 2 \ (C_5H_5)_2TiBH_4 \ + \ 4 \ LiCl \ + \ B_2H_6 \ + \ H_2$$

(purple, volatile at
120°/high vacuum)

These borohydrides have structures **VII** and **VIII** with metal–hydrogen–boron bridges. Trimethylamine reacts with $(C_5H_5)_2Zr(BH_4)_2$ (**VIII**) in benzene

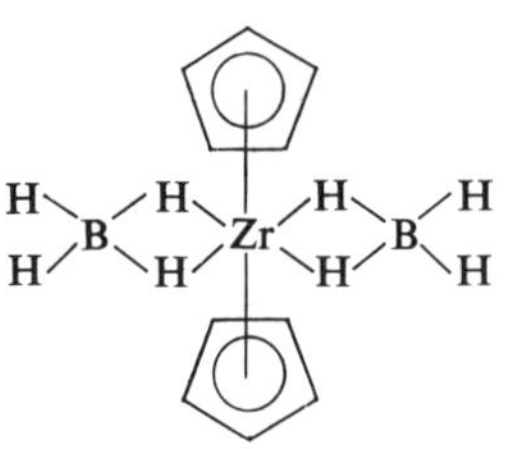

VII 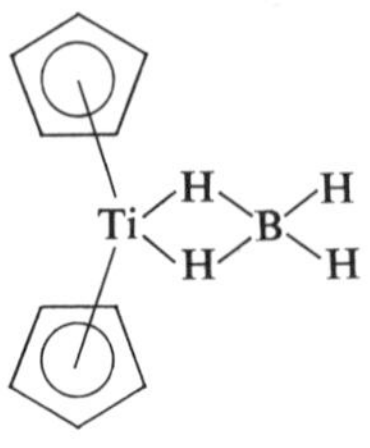   VIII

solution to remove BH_3 groups giving first the white, air-sensitive, volatile mononuclear derivative $(C_5H_5)_2Zr(H)(BH_4)$ (**IX**) and then the white, insoluble, nonvolatile, polymeric derivative $[(C_5H_5)_2ZrH_2]_n$, apparently **X** with zirconium–hydrogen–zirconium bridges [34].

A neutral biscyclopentadienylzirconium $(C_5H_5)_2Zr$ was recently [35] obtained as a purple-black, pyrophoric, diamagnetic solid by reduction of $(C_5H_5)_2ZrCl_2$ with sodium naphthalenide in tetrahydrofuran solution, analogous to the preparation of biscyclopentadienyltitanium described above. The low solubility and low volatility of $(C_5H_5)_2Zr$ suggests that it is a polynuclear species rather than a simple monomer.

No carbonyl derivatives of zirconium of hafnium have been reported. However, some π-allyl derivatives of these two metals have been obtained.

Treatment of the metal tetrahalides with allylmagnesium chloride in diethyl ether solution gives the red, unstable, air-sensitive, volatile (25°/0.0001 mm) tetraallyl derivatives $(C_3H_5)_4M$ (M = Zr or Hf) [36]. The temperature dependence of the proton NMR spectra of the π-allyl ligands in these complexes has been investigated; evidence for motion within the metal–π-allyl bond was found [37].

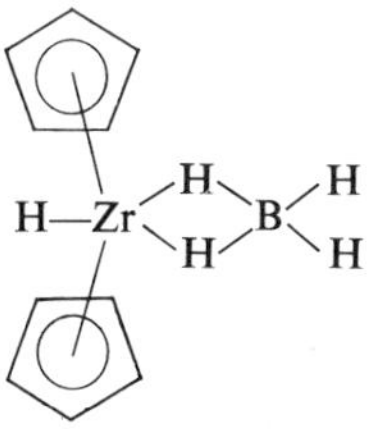

IX 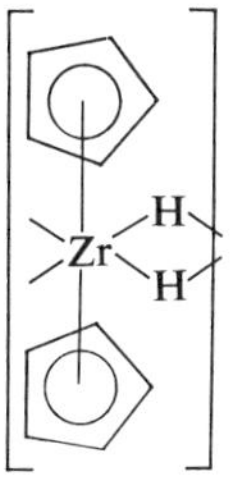   X

Organometallic Chemistry of Vanadium

A neutral vanadium atom requires thirteen electrons from the surrounding ligands to attain the favored eighteen-electron rare-gas electronic configuration. Six electron-pair donors in an octahedral vanadium(0) complex can only supply twelve electrons, giving the vanadium atom in complexes such as hexacarbonylvanadium $V(CO)_6$ only a seventeen-electron configuration, one short of the favored rare-gas electronic configuration. Hexacarbonylvanadium $V(CO)_6$ is known but is extremely unstable compared with the hexacarbonyls of chromium, molybdenum, and tungsten, which have the favored eighteen-electron rare-gas configuration. Hexacarbonylvanadium can easily pick up an additional electron, giving the hexacarbonylvanadate anion $V(CO)_6^-$ which does have the favored eighteen-electron rare-gas electronic configuration and, therefore, is much more stable than neutral $V(CO)_6$. Cyclopentadienyltetracarbonylvanadium $C_5H_5V(CO)_4$ is an example of a neutral vanadium carbonyl derivative with the favored eighteen-electron rare-gas electronic configuration.

The most readily prepared vanadium carbonyl derivative is the hexacarbonylvanadate salt of the bisdiglyme sodium cation, which can be prepared in fairly good yield [3] by treatment of a mixture of vanadium trichloride and metallic sodium in diglyme with carbon monoxide under pressure according to the following equation [38]:

$$VCl_3 + 4\,Na + 2\,C_6H_{14}O_3 + 6\,CO \xrightarrow[\text{48 hr}]{160°/300\ atm} [Na(C_6H_{14}O_3)_2][V(CO)_6] + 3\,NaCl$$

(yellow solid,
light sensitive,
$\nu(CO) = 1860\ cm^{-1}$)

The fairly unusual *cation* in this salt appears to consist of a sodium atom octahedrally coordinated with the six oxygen atoms of the two tridentate ether molecules; this cation has not been encountered very much outside metal carbonyl chemistry. When salts of the $V(CO)_6{}^-$ anion are acidified with concentrated aqueous hydrochloric acid in the presence of a second diethyl ether phase, they give a yellow-orange ethereal layer, apparently containing the oxonium salt $[(C_5H_5)_2OH][V(CO)_6]$. Upon evaporation this oxonium salt decomposes, giving neutral hexacarbonylvanadium $V(CO)_6$, apparently according to the following equation [39]:

$$2\,[(C_2H_5)_2OH][V(CO)_6] \longrightarrow 2\,V(CO)_6 + \tfrac{1}{2}\,H_2 + (C_2H_5)_2O$$

(blue-green,
pyrophoric,
sublimes 25°/0.1 mm,
paramagnetic)

This provides the best available synthesis of neutral hexacarbonylvanadium.

A variety of reactions of hexacarbonylvanadium have been investigated, particularly in the laboratory of Calderazzo (Cyanamid European Research Institute, Geneva, Switzerland). Reaction of hexacarbonylvanadium with triphenylphosphine in hexane solution at 25° gives the orange *trans*-disubstituted derivative $[(C_6H_5)_3P]_2V(CO)_4$ (**XI**) which has the same seventeen-electron configuration as $V(CO)_6$ and, therefore, also is paramagnetic (1.78 BM), corresponding to one unpaired electron [40]. This complex **XI** may, therefore, be considered as an organovanadium free radical, and, as such,

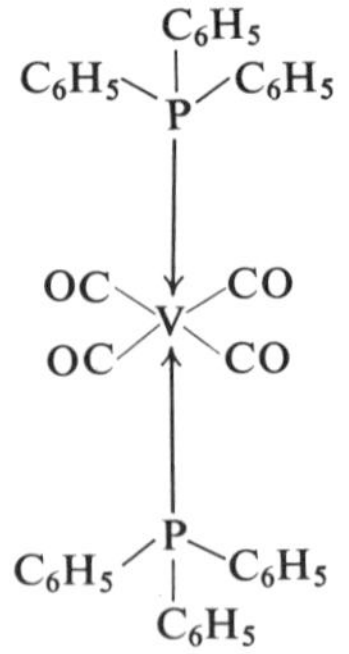

C$_6$H$_5$
C$_6$H$_5$\ | /C$_6$H$_5$
P
|
OC\ ↓ /CO
V
OC/ ↑ \CO
|
C$_6$H$_5$ /P\ C$_6$H$_5$
C$_6$H$_5$

XI

reacts with nitric oxide, an inorganic free radical, to give the orange vanadium nitrosyl derivative $(C_6H_5)_3PV(CO)_4NO$, mp 88°–90°. This nitrosyl derivative is a substitution product of the unknown $V(CO)_5NO$, which has the favored eighteen-electron rare-gas configuration. Solutions of the unsubstituted $V(CO)_5NO$ can be prepared by treatment of solutions of $V(CO)_6$ with nitric oxide at −78°. Attempts to isolate pure $V(CO)_5NO$ from these solutions lead to decomposition; however, the presence of $V(CO)_5NO$ in these solutions could be established by infrared spectroscopy in the $\nu(CO)$ and $\nu(NO)$ regions.

Hexacarbonylvanadium also reacts with certain hydrocarbons. In the case of aromatic hydrocarbons, including benzene and its methylated derivatives, reaction proceeds at 35°–50°, giving 5 to 30% yields of the red salts $[(arene)V(CO)_4][V(CO)_6]$ (**XII**: arene = benzene), according to the following equation [41]:

$$2\ V(CO)_6 + arene \longrightarrow [(arene)V(CO)_4][V(CO)_6] + 2\ CO$$

The occurrence of this disproportionation reaction is a further illustration of the tendency for the seventeen-electron species $V(CO)_6$ to be converted to the more stable eighteen-electron species $V(CO)_6^-$. Other salts (e.g., PF_6^- and $B(C_6H_5)_4^-$ of the $[(arene)V(CO)_4]^+$ cations can be obtained by appropriate metathesis reactions. Furthermore, hydride addition to the $[(arene)V(CO)_4]^+$ cations by means of sodium borohydride in tetrahydrofuran solution gives π-cyclohexadienylvanadium tetracarbonyl derivatives [42]; thus, the benzene derivative $[C_6H_6V(CO)_4]^+$ reacts with sodium borohydride in tetrahydrofuran solution to give the brown, volatile, unsubstituted π-cyclohexadienyl derivative $C_6H_7V(CO)_4$ (**XIII**), mp 66° (dec.). This type of hydride addition to arene–metal cations is a commonly used method for the preparation of π-cyclohexadienyl derivatives.

XII   XIII

The reaction between cycloheptatriene and hexacarbonylvanadium has been found to give two products of interest [43]. The first product is the non-ionic π-cycloheptatrienyl derivative $C_7H_7V(CO)_3$ (**XIV**), a green, volatile, diamagnetic solid. The green color of this compound is typical of neutral

π-cycloheptatrienylmetal derivatives with an eighteen-electron rare-gas configuration. The second product is the brown hexacarbonylvanadate salt $[C_7H_7VC_7H_8][V(CO)_6]$ (XV). The cation of this salt is unique, since it contains both π-cycloheptatrienyl (C_nH_n) and π-cycloheptatriene (C_7H_8) ligands. The π-cycloheptatrienyl ligand has all seven carbon atoms of the seven-membered ring bonded to the metal atom, whereas the π-cycloheptatriene ligand has but six of the carbon atoms of the seven-membered ring bonded to the metal atom. The cation $[C_7H_7VC_7H_8]^+$ has only a seventeen-electron configuration and exhibits the expected paramagnetism for one unpaired electron.

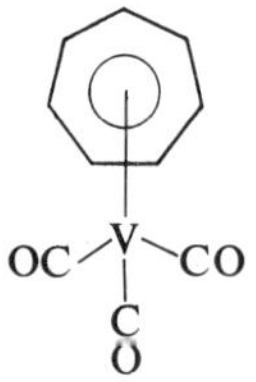

XIV 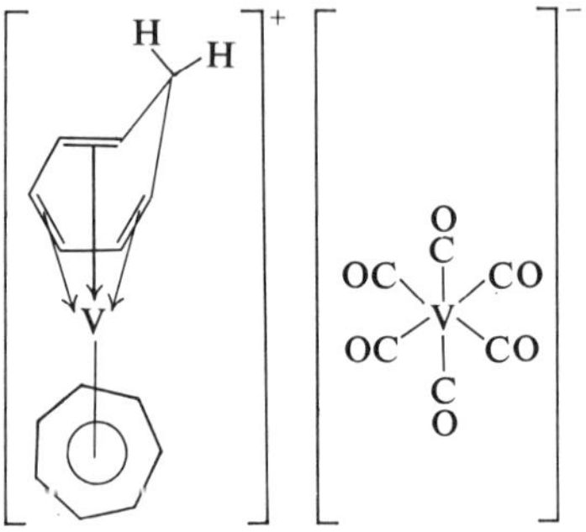   XV

Vanadium forms a variety of π-cyclopentadienyl derivatives. Reaction of vanadium tetrachloride with two equivalents of sodium cyclopentadienide in 1,2-dimethoxyethane solution gives green, air-stable $(C_5H_5)_2VCl_2$ according to the following equation [10b]:

$$VCl_4 + 2\,NaC_5H_5 \longrightarrow (C_5H_5)_2VCl_2 + 2\,NaCl$$

In this compound the vanadium atom remains in the 4+ formal oxidation state. However, treatment of vanadium tetrachloride or vanadium trichloride with excess sodium cyclopentadienide gives biscyclopentadienylvanadium (vanadocene), $(C_5H_5)_2V$, in good yield [44]. Biscyclopentadienylvanadium is a purple, very air-sensitive solid subliming readily at $70°/0.1$ mm and exhibiting the $\sim 170°$ melting point characteristic of the biscyclopentadienyls of the first-row transition metals $(C_5H_5)_2M$ (M = V, Cr, Mn, Fe, Co, or Ni).

Some of the most interesting reactions of $(C_5H_5)_2V$ are those with carbonylating agents. Reaction of $(C_5H_5)_2V$ with hexacarbonylvanadium in the presence of carbon monoxide at atmospheric pressure proceeds according to the following equation [45]:

$$(C_5H_5)_2V + 2\,CO + V(CO)_6 \longrightarrow [(C_5H_5)_2V(CO)_2][V(CO)_6]$$

The product of this reaction is the hexacarbonylvanadate salt of the $[(C_5H_5)_2V(CO)_2]^+$ cation, which is isoelectronic with the neutral titanium carbonyl derivative $(C_5H_5)_2Ti(CO)_2$. Reaction of $(C_5H_5)_2V$ with carbon monoxide under pressure at temperatures above $100°$ removes one π-cyclopentadienyl ring, giving orange cyclopentadienyltetracarbonylvanadium, $C_5H_5V(CO)_4$ (XVI), mp $138°$, subl. $80°/0.1$ mm [44]. This vanadium complex XVI has the eighteen-electron rare-gas configuration and exhibits the expected diamagnetism. A better synthesis of $C_5H_5V(CO)_4$ (XVI) utilizes the following reaction between $[Na(diglyme)_2][V(CO)_6]$ and the cyclopentadienyl-mercuric chloride generated by mixing sodium cyclopentadienide and mercuric chloride in tetrahydrofuran solution in a $1:1$ molar ratio [46]:

$$NaC_5H_5 + HgCl_2 \xrightarrow{\text{THF}} C_5H_5HgCl + NaCl$$

$$[Na(diglyme)_2][V(CO)_6] + C_5H_5HgCl \xrightarrow{\text{THF}} C_5H_5V(CO)_4 + Hg + 2\,CO + diglyme + NaCl$$

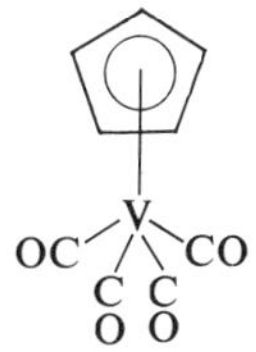

XVI    XVII

A variety of interesting reactions of $C_5H_5V(CO)_4$ (XVI) have been investigated. Ultraviolet irradiation of $C_5H_5V(CO)_4$ with conjugated dienes (e.g., butadiene and 1,3-cyclohexadiene) in hexane solution replaces two of the four carbonyl groups with the diene ligand, giving the red complexes $C_5H_5V(CO)_2(diene)$ [47]. Heating $C_5H_5V(CO)_4$ in boiling cycloheptatriene replaces all four carbonyl groups with a π-cycloheptatrienyl ligand, giving purple, volatile $(100°/0.1$ mm) π-cyclopentadienyl-π-cycloheptatrienyl-vanadium $C_5H_5VC_7H_7$ (XVII) [48]. This complex has only a seventeen-electron configuration for the vanadium atom and exhibits the expected paramagnetism of 1.69 BM for one unpaired electron. Ultraviolet irradiation of $C_5H_5V(CO)_4$ with alkynes results in the replacement of two carbonyl groups with one alkyne ligand, giving green complexes of the type $C_5H_5V(CO)_2(alkyne)$ [49]. These complexes probably have bidentate monometallic alkyne ligands (p. 30).

Some types of oxidizing reagents can oxidize $C_5H_5V(CO)_4$ with formally 1+ vanadium to derivatives with the vanadium atom in higher oxidation states, with simultaneous loss of all four carbonyl groups. Thus, $C_5H_5V(CO)_4$ reacts with dimethyl disulfide at $100°$, with complete loss of carbonyl groups, to give brown, volatile $(\sim150°/0.1$ mm) $(C_5H_5V(SCH_3)_2)_2$ which appears to have structure XVIII with four bridging CH_3S groups [50]. Reaction of

$C_5H_5V(CO)_4$ with hydrogen chloride and oxygen gives the blue volatile oxychloride $C_5H_5VOCl_2$ (**XIX**); this reacts with chlorine to give the trichloride $C_5H_5VCl_3$ [51]. The oxychloride can also be made from $(C_5H_5)_2V$ and air followed by chlorination with thionyl chloride or other good chlorinating agents [52].

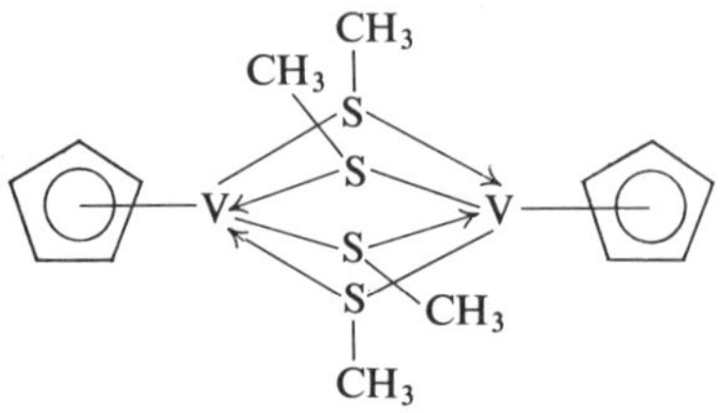

XVIII 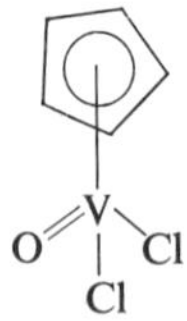   **XIX**

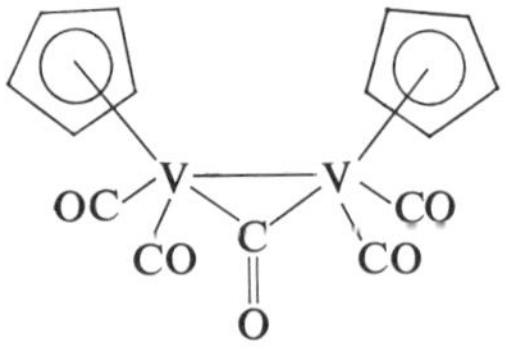

XX

Many metal carbonyl anions can be prepared by reaction of appropriate neutral metal carbonyls with sodium metal. Reaction of $C_5H_5V(CO)_4$ with sodium metal, either dissolved in liquid ammonia or as 1% sodium amalgam in tetrahydrofuran, gives the yellow sodium salt $Na_2[C_5H_5V(CO)_3]$; the anion $[C_5H_5V(CO)_3]^{2-}$ is isoelectronic with the species $[C_5H_5Cr(CO)_3]^-$ and $C_5H_5Mn(CO)_3$ to be discussed in later sections [44, 53]. Acidification of aqueous solutions of the $[C_5H_5V(CO)_3]^{2-}$ does not give a carbonyl hydride derivative as is the case with many other metal carbonyl anions. Instead, the green dinuclear complex $(C_5H_5)_2V_2(CO)_5$ (**XX**) is formed. This green complex **XX** is very reactive. Thus, it is cleaved by tertiary phosphines at room temperature, according to the following equation:

$$(C_5H_5)_2V_2(CO)_5 + 3\ R_3P \longrightarrow C_5H_5V(CO)_3PR_3 + C_5H_5V(CO)_2(PR_3)_2$$

Similar tertiary phosphine substitution products of $C_5H_5V(CO)_4$ can be made by direct reactions of $C_5H_5V(CO)_4$ with the tertiary phosphine, either with ultraviolet irradiation or at elevated temperatures.

In this chapter the vanadium sandwich compounds $(C_5H_5)_2V$, $[C_7H_7VC_7H_8]^+$, and $C_5H_5VC_7H_7$ (**XVII**) have already been discussed. A further representative is dibenzenevanadium, $(C_6H_6)_2V$ (**XXI**), a brown, volatile crystalline solid, mp 277°–278°, best prepared by heating a mixture of vanadium tetrachloride, aluminum chloride, and powdered aluminum in boiling benzene, followed by hydrolysis of the reaction mixture with concentrated aqueous potassium hydroxide [54]. The vanadium atom in dibenzenevanadium (**XXI**) has only a seventeen-electron configuration, thereby accounting for its paramagnetism of 1.73 BM, corresponding to one unpaired electron.

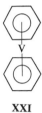

XXI

Organometallic Chemistry of Niobium and Tantalum

The organometallic chemistry of these two metals is relatively limited. Their carbonyl chemistry can be summarized by the following reactions:

$$MCl_5 + 6\ Na + 2\ C_6H_{14}O_3 + 6\ CO \xrightarrow[\substack{200\ atm,\\ Fe\ catalyst}]{100°} [Na(C_6H_{14}O_3)_2]\ [M(CO)_6] + 5\ NaCl$$

(M = Nb or Ta) (yellow, more air-sensitive

than V analogue) [46] (a)

$$[Na(C_6H_{14}O_3)_2][M(CO)_6] + C_5H_5HgCl \xrightarrow{THF} C_5H_5M(CO)_4 + 2\ CO + Hg +$$

(M = Nb or Ta) (orange,   NaCl + 2 C$_6$H$_{14}$O$_3$

sublimes [46] (b)

100°/0.1 mm)

$$[Na(C_6H_{14}O_3)_2][Ta(CO)_6] + RHgCl \longrightarrow RHgTa(CO)_6 + NaCl + 2\ C_6H_{14}O_3 \quad [52]$$

(red, volatile (c)

nonconductor)

(R = ethyl; methyl and phenyl derivatives are less stable)

Equation (a) given above corresponds to an efficient preparation of $V(CO)_6^-$ (p. 53). However, the preparation of the niobium and tantalum derivatives is much more difficult to carry out than that of the vanadium derivative, requiring an iron carbonyl or iron chloride catalyst and more carefully controlled reaction conditions and experimental techniques. Equation (b) corresponds to an efficient preparation of $C_5H_5V(CO)_4$. The niobium compound $C_5H_5Nb(CO)_4$ has also been obtained in trace quantities by carbonylation at elevated temperatures and pressures of an ill-defined reaction mixture obtained from niobium pentachloride, sodium cyclopentadienide, and strong hydridic reducing agents [55]. Equation (c) has no counterpart in vanadium chemistry. The organomercury derivatives of tantalum carbonyl thus obtained are thermally unstable, decomposing slightly above room temperature. These $RHgTa(CO)_6$ compounds appear to be derivatives of seven-coordinate tantalum. No neutral derivatives of niobium and tantalum containing just carbonyl ligands [cf. $V(CO)_6$] are known.

Other known π-cyclopentadienyl derivatives of niobium and tantalum include the halides $(C_5H_5)_2MX_3$ (M = Nb or Ta; X = Cl, Br, or I), the trihydride $(C_5H_5)_2TaH_3$, the borohydride $(C_5H_5)_2NbCl(BH_4)$, and the tetra-cyclopentadienyls $(C_5H_5)_4M$ (M = Nb or Ta). A green-black tetraallylniobium, $(C_3H_5)_4Nb$, has been obtained by treatment of niobium pentachloride with excess allylmagnesium chloride. This compound is very thermally unstable, decomposing above 0° [36]. Hexamethylbenzene derivatives of niobium and tantalum have been obtained by fusing the metal pentachlorides with hexamethylbenzene, aluminum powder, and aluminum chloride at 130°, followed by hydrolysis to give brown nonionic complexes of the type $[(CH_3)_6C_6MCl_2]_2$ and green ionic derivatives of the type $[(CH_3)_6C_6MCl_2]_3{}^+Cl^-$. No bis(hexamethylbenzene)metal derivatives of niobium and tantalum with two rings per metal atom are known [56].

A methylniobium compound with a niobium–carbon σ-bond has been prepared [57]. Reaction of niobium pentachloride with dimethylzinc in pentane solution gives golden yellow, volatile (25°/0.001 mm) trimethyl-niobium dichloride $(CH_3)_3NbCl_2$, according to the following equation:

$$2\ NbCl_5 + 3\ (CH_3)_2Zn \longrightarrow 2\ (CH_3)_3NbCl_2 + 3\ ZnCl_2$$

Trimethylniobium dichloride is stable at $-78°$ but decomposes upon heating to 25° or in the presence of water or oxygen. An analogous but less stable trimethyltantalum dichloride is also known.

REFERENCES

1. J. M. Birmingham and G. Wilkinson, *J. Am. Chem. Soc.* **78**, 42 (1956).
2. E. O. Fischer and H. Fischer, *J. Organometal. Chem. (Amsterdam)* **6**, 141 (1966).
3. F. Calderazzo, R. Pappalardo, and S. Cosi, *J. Inorg. & Nucl. Chem.* **28**, 987 (1966).
4. E. O. Fischer and H. Fischer, *J. Organometal. Chem. (Amsterdam)* **3**, 181 (1965).
5. E. O. Fischer and A. Treiber, *Z. Naturforsch.* **17b**, 276 (1962).
6. G. L. Ter Haar and M. Dubeck, *Inorg. Chem.* **3**, 1648 (1964).
7. E. O. Fischer and Y. Hristidu, *Z. Naturforsch.* **17b**, 275 (1962).
8. G. L. Ter Haar and M. Dubeck, *Inorg. Chem.* **3**, 1648 (1964).
9a. E. O. Fischer, P. Laubereau, F. Baumgärtner, and B. Kanellakopulos, *J. Organometal. Chem. (Amsterdam)* **5**, 583 (1966).
9b. E. O. Fischer, P. Laubereau, F. Baumgärtner, and B. Kanellakopulos, *Angew. Chem.* **77**, 866 (1965).
9c. E. O. Fischer, P. Laubereau, F. Baumgärtner, and B. Kanellakopulos, *Angew. Chem. Intern. Ed. Engl.* **5**, 134 (1966).
10a. A. Streitwieser, Jr. and U. Müller-Westerhoff, *J. Am. Chem. Soc.* **90**, 7364 (1968).

10b. G. Wilkinson and J. M. Birmingham, *J. Am. Chem. Soc.* **76**, 4281 (1954).
11. L. Summers, R. H. Uloth, and A. Holmes, *J. Am. Chem. Soc.* **77**, 3604 (1955).
12. J. G. Murray, *J. Am. Chem. Soc.* **83**, 1287 (1961).
13. H. A. Martin and F. Jellinek, *Angew. Chem.* **76**, 274 (1964); *J. Organometal. Chem.* (*Amsterdam*) **6**, 293 (1966).
14. R. D. Gorsich, *J. Am. Chem. Soc.* **82**, 4211 (1960).
15. A. N. Nesmeyanov, O. V. Nogina, and A. M. Berlin, *Dokl. Akad. Nauk SSSR* **134**, 607 (1960).
16. U. Giannini and S. Cesca, *Tetrahedron Letters* p. 19 (1960).
17. G. A. Razuvaev, V. N. Latyaeva, L. I. Vyshinskaya, and N. N. Vyshinskii, *Dokl. Akad. Nauk SSSR* **15b**, 1121 (1964).
18. E. O. Fischer and A. Löchner, *Z. Naturforsch.* **15b**, 266 (1960).
19. S. A. Giddings, *Inorg. Chem.* **3**, 684 (1964); A. F. Reid and P. C. Wailes, *Australian J. Chem.* **18**, 9 (1965).
20. P. D. Bartlett and B. Seidel, *J. Am. Chem. Soc.* **83**, 581 (1961).
21. A. K. Fischer and G. Wilkinson, *J. Inorg. & Nucl. Chem.* **2**, 149 (1956).
22. G. W. Watt, L. J. Baye, and F. O. Drummond, *J. Am. Chem. Soc.* **88**, 1138 (1966).
23. M. E. Vol'pin and V. B. Shur, *Nature* **209**, 1236 (1966).
24. H. Brintzinger, *J. Am. Chem. Soc.* **88**, 4305 (1966).
24a. E. E. van Tamelen, R. B. Fechter, S. W. Schneller, G. Boche, R. H. Greeley, and B. Åkermark, *J. Am. Chem. Soc.* **91**, 1551 (1969).
25. C. Beerman and H. Bestian, *Angew. Chem.* **71**, 618 (1959).
26. H. J. Berthold and G. Groh, *Z. Anorg. Allgem. Chem.* **319**, 230 (1963).
27. K. H. Thiele and J. Müller, *Z. Chem.* **4**, 273 (1964).
28. V. N. Latyaeva, G. A. Razuvaev, A. V. Malisheva, and G. A. Kilyakova, *J. Organometal. Chem.* (*Amsterdam*) **2**, 388 (1964).
29. H. Breil and G. Wilke, *Angew. Chem. Intern. Ed. Engl.* **5**, 898 (1966).
30. H. Dietrich and H. Dierks, *Angew. Chem. Intern. Ed. Engl.* **5**, 899 (1966).
31. E. M. Brainina, M. Kh. Minacheva, and R. Kh. Freidlina, *Izv. Akad. Nauk SSSR, Ser. Khim.* p. 1877 (1965).
32. R. K. Nanda and M. G. H. Wallbridge, *Inorg. Chem.* **3**, 1798 (1964).
33. H. Nöth and R. Hartwimmer, *Ber.* **93**, 2238 (1960).
34. B. D. James, R. K. Nanda, and M. G. H. Wallbridge, *Chem. Commun.* p. 849 (1966).
35. G. W. Watt and F. O. Drummond, *J. Am. Chem. Soc.* **88**, 5926 (1966).
36. G. Wilke *et al.*, *Angew. Chem. Intern. Ed. Engl.* **5**, 151 (1966).
37. J. K. Becconsall, B. E. Job, and S. O'Brien, *J. Chem. Soc.*, *A* p. 423 (1967).
38. R. P. M. Werner and H. E. Podall, *Chem. & Ind.* (*London*) p. 144 (1961).
39. R. Ercoli, F. Calderazzo, and A. Alberola, *J. Am. Chem. Soc.* **82**, 2966 (1960); F. Calderazzo and R. Ercoli, *Chim. Ind.* (*Milan*) **44**, 990 (1962).
40. R. P. M. Werner, *Z. Naturforsch.* **16b**, 499 (1961).
41. F. Calderazzo, *Inorg. Chem.* **3**, 1207 (1964).
42. F. Calderazzo, *Inorg. Chem.* **5**, 429 (1966).
43. R. P. M. Werner and S. A. Manastyrskyj, *J. Am. Chem. Soc.* **83**, 2023 (1961); F. Calderazzo and P. L. Calvi, *Chim. Ind.* (*Milan*) **44**, 1217 (1962).
44. E. O. Fischer and S. Vigoureux, *Ber.* **91**, 2205 (1958).
45. F. Calderazzo and S. Bacciarelli, *Inorg. Chem.* **2**, 721 (1963).
46. R. P. M. Werner, A. H. Filbey, and S. A. Manastyrskyj, *Inorg. Chem.* **3**, 298 (1964).
47. E. O. Fischer, H. P. Kögler, and P. Kuzel, *Ber.* **93**, 3006 (1960).
48. R. B. King and F. G. A. Stone, *J. Am. Chem. Soc.* **81**, 5263 (1959).
49. R. Tsumura and N. Hagihara, *Bull. Chem. Soc. Japan* **38**, 1901 (1965).

50. R. H. Holm, R. B. King, and F. G. A. Stone, *Inorg. Chem.* **2,** 219 (1963).
51. E. O. Fischer, S. Vigoureux, and P. Kuzel, *Ber.* **93,** 701 (1960).
52. H. J. de Liefde Meijer and G. J. M. Van der Kerk, *Rec. Trav. Chim.* **84,** 1418 (1965).
53. E. O. Fischer and R. J. J. Schneider, *Angew. Chem. Intern. Ed. Engl.* **6,** 569 (1967).
54. E. O. Fischer and H. P. Kögler, *Ber.* **90,** 250 (1957).
55. R. B. King, *Z. Naturforsch.* **18b,** 157 (1963).
56. E. O. Fischer and R. Röhrscheid, *J. Organometal. Chem.* (*Amsterdam*) **6,** 53 (1966).
57. G. L. Juvinall, *J. Am. Chem. Soc.* **86,** 4202 (1964).

SUPPLEMENTARY READING

H. Bestian and K. Clauss, Reactions of olefins with the titanium-carbon bond. *Angew. Chem. Intern. Ed. Engl.* **2,** 704 (1963).

QUESTIONS

1. Discuss possibilities for preparing anionic carbonyl derivatives of titanium.
2. Discuss the magnetic properties of π-cyclopentadienyltitanium derivatives.
3. Suggest a structure for the tetracyclopentadienyls of zirconium and hafnium.
4. Suggest preparative applications of the cyclooctatetraene dianion ($C_8H_8^{2-}$) in organo-titanium chemistry.
5. Suggest possible extensions of the carbonyl chemistry of niobium and tantalum.

CHAPTER III

Organometallic Derivatives of Chromium, Molybdenum, and Tungsten

Introduction

Neutral chromium, molybdenum, and tungsten atoms each require twelve electrons from the surrounding ligands to attain the eighteen-electron configuration of the next rare gas. These twelve electrons can be supplied by six monodentate ligands in an octahedral zerovalent complex. Thus, the hexacarbonyls of chromium, molybdenum, and tungsten, $M(CO)_6$, have the favored rare-gas configurations and, therefore, are stable compounds. Alternatively, the twelve electrons required to attain a rare-gas configuration can be supplied by two benzene rings in a dibenzene–metal complex, making dibenzenechromium and corresponding derivatives of molybdenum and tungsten stable compounds. Neutral cyclopentadienylmetal (C_5H_5M) fragments of chromium, molybdenum, and tungsten each require seven more electrons from other ligands to attain the eighteen-electron configuration of the next rare gas. Cyclopentadienylmetal carbonyl derivatives of chromium, molybdenum, and tungsten of the types $C_5H_5M(CO)_3X$ and $C_5H_5M(CO)_2T$ ($M = Cr$, Mo, or W; $X =$ one-electron donor; $T =$ three-electron donor) are thus stable compounds.

The Metal Hexacarbonyls

All three metals form white, air-stable, diamagnetic, odorless, crystalline hexacarbonyls of the type $M(CO)_6$ ($M = Cr$, Mo, or W), which have an octahedral configuration of the six carbonyl groups about the central metal atom. The volatility of these metal hexacarbonyls is sufficiently high so that they evaporate upon standing in air, sublime easily at room temperature, and can be purified by steam distillation. These metal hexacarbonyls are prepared by reductive carbonylation reactions such as the following, all of which are best carried out under carbon monoxide pressures of 100 to 300 atm.

HEXACARBONYLCHROMIUM

$$CrCl_3 + RMgX + CO \xrightarrow{\text{ether}} \xrightarrow{H_3O^+} Cr(CO)_6 + \text{other products} \quad [1] \quad (a)$$

$$Cr(C_5H_7O_2)_3 + Mg + CO \xrightarrow[I_2]{\text{pyridine}} \xrightarrow{H_3O^+} Cr(CO)_6 + \text{other products} \quad [2] \quad (b)$$

$$CrCl_3 + AlCl_3 + Al + CO \xrightarrow{\text{benzene}} \xrightarrow{H_3O^+} Cr(CO)_6 + \text{other products} \quad [3] \quad (c)$$

Reaction (a), using an organomagnesium compound (Grignard reagent) as the reducing agent, is now mainly of historical interest as representing the first successful preparation of hexacarbonylchromium. Reactions (b) and (c) represent practical methods for the preparation of hexacarbonylchromium.

HEXACARBONYLS OF MOLYBDENUM AND TUNGSTEN

$$MCl_x + Zn + CO \xrightarrow{\text{ether}} \xrightarrow{H_3O^+} M(CO)_6 + \text{other products} \quad [4] \quad (a)$$

$$MCl_x + Fe(CO)_5 \xrightarrow[H_2]{\text{ether}} \xrightarrow{H_3O^+} M(CO)_6 + \text{other products} \quad [5] \quad (b)$$

where M = Mo and X = 5 or M = W and X = 6.

Reaction (b) is of interest in illustrating the use of pentacarbonyliron as a source of carbon monoxide in the preparation of metal carbonyls. Reaction (a) illustrates a practical method for the preparation of the hexacarbonyls of molybdenum and tungsten.

Reactions of the Metal Hexacarbonyls

The metal hexacarbonyls are extremely useful intermediates for the preparation of organometallic compounds of chromium, molybdenum, and tungsten. Most of the reactions of the metal hexacarbonyls involve the nucleophilic replacement of one or more carbonyl groups with other ligands. In most cases replacement of some of the carbonyl groups in the metal hexacarbonyls markedly increases the resistance of the remaining carbonyl groups towards substitution. Thus, it is often relatively easy to control the degree of substitution of the metal hexacarbonyls by controlling the reaction conditions. Furthermore, it is very difficult to replace all six carbonyl groups of the metal hexacarbonyls with other ligands. Indeed, many ligands including pyridine and acetonitrile do not appear to be capable of replacing more than three carbonyl groups in the metal hexacarbonyls, even under vigorous reaction conditions. A further feature of the reactivity of the metal hexacarbonyls is that hexacarbonyltungsten is significantly less reactive than hexacarbonyl-molybdenum or hexacarbonylchromium. This effect is sufficiently large to make many reactions suitable for the preparation of molybdenum and chromium carbonyl derivatives no longer suitable for the preparation of analogous tungsten carbonyl derivatives.

π-BONDING

Explanation of some of these observations on metal hexacarbonyl reactions has generated some controversy. The π-bonding school of thought considers a major contributor to the strength of the metal–carbon bond to be retrodative π-bonding, which increases as the negative charge on the metal atom increases. Replacement of a carbonyl group with a more weakly π-accepting ligand (apparently all commonly encountered ligands except PF_3) increases the electron density on the metal atom. More electrons are thus available for retrodative π-bonding with the remaining carbonyl groups. This additional retrodative π-bonding strengthens the metal–carbon bond but weakens the carbon–oxygen bond, since increased retrodative π-bonding increases the occupation of the antibonding orbitals of the carbon–oxygen bond. Changes in the carbon–oxygen bond order in carbonyl groups arising from changes in retrodative π-bonding can be determined by investigation of the $\nu(CO)$ frequencies in the infrared spectrum in the range 2100–1700 cm^{-1}. A decrease in $\nu(CO)$ frequency can be taken as an indication of increased retrodative π-bonding.

σ-BONDING

Other workers attribute these observations concerning the chemistry of octahedral metal carbonyl derivatives to changes in σ-bonding. Replacement of carbonyl groups in octahedral metal carbonyl derivatives with more strongly basic ligands (the same ligands as the more weakly π-accepting ligands mentioned above) makes the substituted metal atom a weaker Lewis acid less likely to form complexes with Lewis bases.

Metal Carbonyl Halide Derivatives

Halide ions (except fluoride) behave as Lewis bases toward the metal hexacarbonyl, replacing one of the six carbonyl groups to form halopenta-carbonyl metallates according to the following equation [6, 7]:

$$M(CO)_6 + M'X \xrightarrow[65°-120°]{ethers} M'[M(CO)_5X] + CO$$

where $M = Cr$, Mo, or W; $M' = N$-methylpyridinium or tetraalkyl-ammonium; ether solvents: tetrahydrofuran, dioxane, or diglyme; $X = Cl$, Br, I, or NCS. A nonhydroxylic but coordinating solvent appears to be necessary for this reaction. Alkali-metal halides also react with the metal hexacarbonyls in this manner, but pure alkali-metal halopentacarbonyl-metallate salts cannot be isolated.

The halopentacarbonylmetallates are yellow solids which can be handled briefly in air but which oxidize in solution or on prolonged exposure to air. They undergo a variety of reactions of interest. Thus, the eighteen-electron anion $[Cr(CO)_5I]^-$ can be oxidized by Fe^{3+} or I_3^- to the seventeen-electron, neutral, deep blue $Cr(CO)_5I$, which exhibits the expected paramagnetism for one unpaired electron [8]:

$$Cr(CO)_5I^- + Fe^{3+} \longrightarrow Cr(CO)_5I + Fe^{2+}$$

This complex $Cr(CO)_5I$ is one of the best-known octahedral derivatives of chromium(I). It is reasonably stable to air oxidation over short periods of time, but does undergo thermal decomposition at temperatures only slightly above room temperature. The molybdenum and tungsten anions of the type $[M(CO)_5X]^-$ $(X = Cl, Br, or I)$ do not undergo a similar oxidation to a neutral seventeen-electron complex, but, instead, react with free halogens to form a heptacoordinate metal(II)–tetracarbonyltrihalide complex according to the following equation [9]:

$$M(CO)_5X^- + Y_2 \longrightarrow [M(CO)_4XY_2]^- + CO$$
$$\text{(yellow)}$$

The smaller chromium atom does not appear to form analogous hepta-coordinate complexes. The halide ligands in the halopentacarbonylmetallates, $M(CO)_5X^-$, can be replaced by neutral Lewis bases such as amines and isocyanides according to the following equation [10]:

$$M(CO)_5X^- + L \longrightarrow LM(CO)_5 + X^-$$

Sometimes the ligands L will react further with the $LM(CO)_5$ products replac-ing additional carbonyl groups forming $L_2M(CO)_4$ or $L_3M(CO)_3$ compounds. The halopentacarbonylmetallates $M(CO)_5X^-$ react with allyl halides C_3H_5Y to form binuclear derivatives $[M_2Y_3(C_3H_5)_2(CO)_4]^-$ according to the following equation [11]:

$$2 M(CO)_5X^- + 3 C_3H_5Y \longrightarrow [M_2Y_3(C_3H_5)_2(CO)_4]^- + X^- + C_3H_5X + 6 CO$$

These binuclear anions appear to have structure I with three bridging halides and one π-allyl ligand bonded to each metal atom.

Some neutral metal carbonyl halides of molybdenum and tungsten are also known. Treatment of the hexacarbonyls with chlorine or bromine at $-78°$

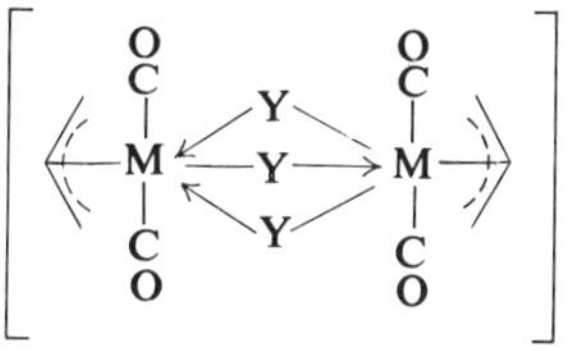

$$\left[\begin{array}{c} \text{O} \\ \text{C} \\ \text{M} \\ \text{C} \\ \text{O} \end{array} \quad \begin{array}{c} \text{Y} \\ \text{Y} \\ \text{Y} \end{array} \quad \begin{array}{c} \text{O} \\ \text{C} \\ \text{M} \\ \text{C} \\ \text{O} \end{array} \right]$$

I

forms the yellow carbonyl halides $[M(CO)_4X_2]_2$ according to the following equation [12]:

$$2\,M(CO)_6 + 2\,X_2 \xrightarrow{\;-78°\;} [M(CO)_4X_2]_2 + 4\,CO$$
$$\text{(yellow)}$$

where $M = Mo$ or W; $X = Cl$ or Br. The binuclear complexes appear to have structure **II** with heptacoordinate metal atoms and two halogen bridges. Ligands such as triphenylphosphine and pyridine break the halogen bridges forming deep yellow monometallic heptacoordinate complexes of the type $L_2Mo(CO)_3X$. Again, the smaller chromium atom does not appear to form analogous heptacoordinate metal carbonyl halide derivatives.

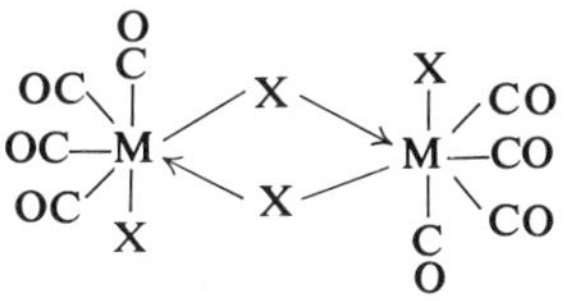

II

Metal Carbonyl Anions

Hexacarbonylchromium may be converted to the anion $Cr(CO)_5^{2-}$ by reduction with sodium metal in liquid ammonia [13]. The yellow air-sensitive sodium salt of this anion may be isolated. The anion $Cr(CO)_5^{2-}$ has a trigonal bipyramidal structure. It reacts with water to form initially the octahedral protonated $HCr(CO)_5{}^-$, which may either decompose further to $Cr(CO)_6$ and hydrogen, or which may react with Lewis bases around room temperature to form $LCr(CO)_5$ complexes. Reaction of the metal hexacarbonyls with sodium borohydride in boiling tetrahydrofuran gives the anions $[HM_2(CO)_{10}]^-$ which may be isolated as tetraethylammonium salts [14]. Crystal-structure study of this anion by x-ray diffraction indicates structure **III**, with a hydrogen atom bridging two $M(CO)_5$ groups [15]. The NMR spectrum of this bridging hydrogen atom exhibits a very high field resonance above τ 20, similar to the high field resonances observed in other transition-metal hydrides. The proton in the $[HM_2(CO)_{10}]^-$ anions can be removed with sodium amalgam to form the yellow $[M_2(CO)_{10}]^{2-}$ anions; the two $M(CO)_5$ groups in these anions are

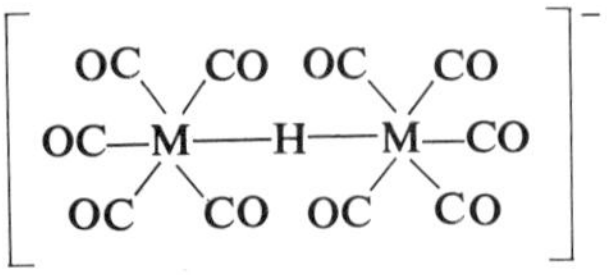

$$
\left[
\begin{array}{ccc}
OC & CO\ \ OC & CO \\
\diagdown\diagup & \diagdown\diagup & \\
OC\!-\!M\!-\!\!-\!H\!-\!\!-\!M\!-\!CO \\
\diagup\diagdown & \diagup\diagdown & \\
OC & CO\ \ OC & CO
\end{array}
\right]^-
$$

III

linked with metal–metal bonds. Other more complex and less understood metal carbonyl anions appear to be present in solutions of the metal hexa-carbonyls after treatment with strong reducing agents such as alkali metals or alkali metal borohydrides. By contrast, treatment of the metal hexacarbonyls with metal triborohydrides, MB_3H_8, gives reasonably stable yellow salts of the anions $[M(CO)_4B_3H_8]^-$ of structure **IV** containing a novel chelating B_3H_8 ligand [16].

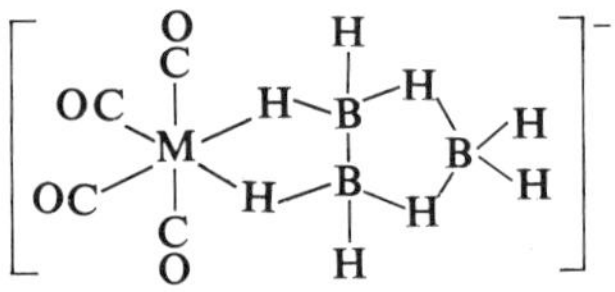

IV

Metal Carbonyl Derivatives of Olefins, Alkynes, and Aromatic Compounds (Arenes)

The metal hexacarbonyls and their derivatives form a variety of substituted metal carbonyl derivatives with arenes or olefins upon heating or ultraviolet irradiation. Especially characteristic are the reactions of hexacarbonyl-chromium with various arenes at temperatures from 110° to 220° to form arenetricarbonylchromium complexes, e.g., $C_6H_6Cr(CO)_3$ (**V**) from benzene

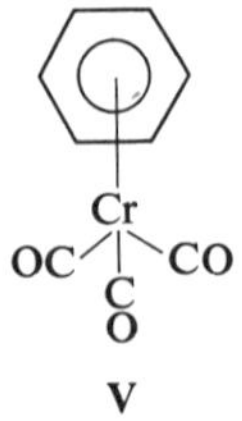

V

and $Cr(CO)_6$ [17–19]. The variety of arenes forming arenetricarbonyl-chromium complexes upon heating with $Cr(CO)_6$ is large. Benzene, toluene, all three xylenes, mesitylene, durene, hexamethylbenzene, aniline, anisole, and chlorobenzene all form yellow arenetricarbonylchromium complexes. Benz-aldehyde, benzoic acid, and nitrobenzene do *not* form arenetricarbonyl-chromium complexes upon heating with hexacarbonylchromium. However, the benzoic acid–chromium tricarbonyl complex $(C_6H_5COOH)Cr(CO)_3$ can be prepared by alkaline hydrolysis followed by acidification of the orange-red methyl benzoate chromium tricarbonyl $(C_6H_5CO_2CH_3)Cr(CO)_3$ obtained by heating $Cr(CO)_6$ with methyl benzoate. Similarly, benzaldehydetricarbonyl-chromium $(C_6H_5CHO)Cr(CO)_3$ can be made by acid cleavage of the chromium tricarbonyl complex of benzaldehyde diethylacetal $[C_6H_5CH(OC_2H_5)_2]Cr(CO)_3$, which can be made by heating the free acetal with $Cr(CO)_6$. Nitrobenzenetri-carbonylchromium has not yet been prepared; the compatibility of the strongly oxidizing nitro group and the arene–chromium tricarbonyl system appears to be doubtful. Furthermore, electronegative substituents such as nitro groups or halogen atoms inhibit formation of chromium tricarbonyl complexes by arenes, apparently because of withdrawing electrons from the bonding orbitals of the arenes. For this reason hexafluorobenzene and hexachloro-benzene do not form chromium tricarbonyl complexes. The polycyclic aro-matic hydrocarbons naphthalene (orange), phenanthrene (orange), anthracene (purple), and pyrene (red) form chromium tricarbonyl complexes of the indicated colors [20]. Thiophene reacts with hexacarbonylchromium to form orange $C_4H_4SCr(CO)_3$ [21]. However, much better yields of the thiophene complex $C_4H_4SCr(CO)_3$ can be obtained by reaction of thiophene with a mixture of (α-picoline)$_3$Cr(CO)$_3$ or (pyridine)$_3$Cr(CO)$_3$ and boron trifluoride at room temperature. Similar systems are useful for preparing arene–chromium tricarbonyls under mild conditions [21].

Molybdenum and tungsten also form arene–metal carbonyl complexes, but with a much more limited series of arenes, mainly benzene and its methylated derivatives. For the syntheses of some of the tungsten derivatives, the aceto-nitrile complex $(CH_3CN)_3W(CO)_3$ (obtained by boiling hexacarbonyltungsten and acetonitrile for 5 days [22]) is more advantageously used for reaction with the arene than $W(CO)_6$, in view of the greater reactivity of the acetonitrile complex than unsubstituted $W(CO)_6$.

Various diolefins and triolefins form complexes upon reacting with either the hexacarbonyls $M(CO)_6$ (M = Cr, Mo, or W), the diglyme complex $(CH_3OCH_2CH_2OCH_2CH_2OCH_3)Mo(CO)_3$, or the acetonitrile complexes $(CH_3CN)_3M(CO)_3$ (M = Cr, Mo, or W). The diglyme and acetonitrile complexes require milder conditions (30°–80°) for reactions with the diolefins and triolefins than the unsubstituted hexacarbonyls (100°–175°) and, therefore, are useful for preparing the less thermally stable complexes such as those of cyclooctatetraene. The reactivity of hexacarbonyltungsten towards diolefins and triolefins is sufficiently low that it is almost always better to use the acetonitrile complex $(CH_3CN)_3W(CO)_3$ for the preparation of tungsten carbonyl complexes of diolefins and triolefins [23].

The following exemplify the types of complexes which have been prepared from carbonyl derivatives of chromium, molybdenum, or tungsten and diolefins or triolefins. (a) Triolefins: (triene)$M(CO)_3$: → triene = cycloheptatriene, 1,3,5-cyclooctatriene, 5,6-dimethylenebicycloheptene-2, cyclooctatetraene (only three of four double bonds complexed with the metal atom). (b) Conjugated diolefins: (diene)$_2M(CO)_2$: → diene = 1,3-cyclohexadiene, butadiene, tetraphenylcyclopentadienone. (c) Nonconjugated diolefins: (diene)$M(CO)_4$: → diene = 1,5-cyclooctadiene, norbornadiene, substituted bicyclo[2,2,2]octatrienes (two of three double bonds complexed with metal atom), dicyclopentadiene.

In addition methyl vinyl ketone (2-butenone) reacts with the acetonitrile complexes $(CH_3CN)_3M(CO)_3$ (M = Mo or W) to form the unusual carbonyl-free complexes $(CH_3COCH{=}CH_2)_3M$ (M = W, 60% yield; M = Mo, 2% yield) of presumed structure VI [23].

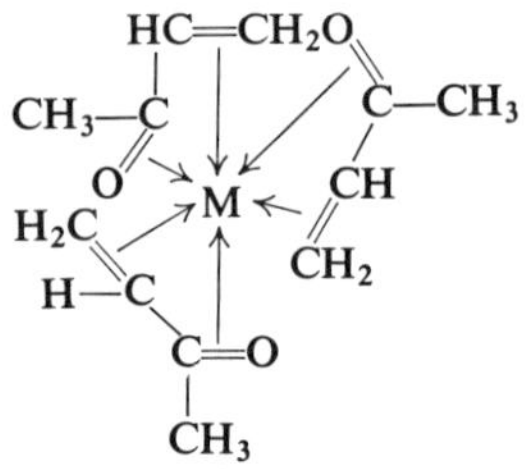

VI

The metal tricarbonyl complexes of cycloheptatriene of formula $C_7H_8M(CO)_3$ (M = Cr, Mo, or W) can be converted to some interesting π-cycloheptatrienyl derivatives. Reaction of the $C_7H_8M(CO)_3$ compounds with triphenylmethyl salts results in hydride abstraction, giving salts of the $[C_7H_7M(CO)_3]^+$ cations, e.g. [24],

$$C_7H_8M(CO)_3 + [(C_6H_5)_3C][PF_6] \xrightarrow[\text{or CH}_3\text{CN}]{\text{CH}_2\text{Cl}_2} [C_7H_7M(CO)_3][PF_6] + (C_6H_5)_3CH$$

(orange)

where M = Cr, Mo, or W. The $[C_7H_7Cr(CO)_3]^+$ cation (and to a lesser extent its molybdenum and tungsten analogues) reacts with some nucleophiles Y^- to give exo-substituted cycloheptatrienetricarbonylchromium derivatives, $C_7H_7YCr(CO)_3$ (VII) [25]. However, with other nucleophiles (e.g., cyclopentadienide) the $[C_7H_7Cr(CO)_3]^+$ cation either undergoes ring contraction, forming the benzene derivative $C_6H_6Cr(CO)_3$ (**V**) [26], or coupling, forming the ditropyl (bicycloheptatrienyl) derivative $(CO)_3CrC_7H_7$—$C_7H_7Cr(CO)_3$ [27]. The $[C_7H_7M(CO)_3]^+$ (M = Mo or W) cations react with halide ions in acetone solution with carbon monoxide evolution to form the green $C_7H_7M(CO)_2X$ (**VIII**) derivatives, i.e. [28],

$$[C_7H_7M(CO)_3][BF_4] + NaX \xrightarrow{\text{acetone}} C_7H_7M(CO)_2X + CO + NaBF_4$$
$$\text{(deep green)}$$

where M = Mo or W; X = Cl, Br, or I.

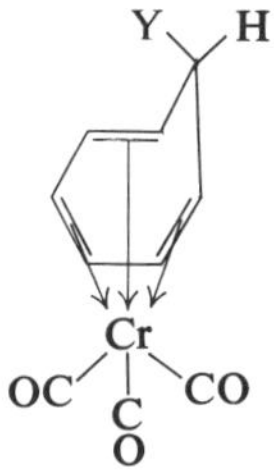

VII 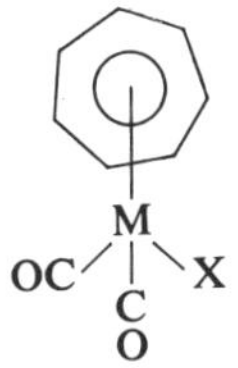   VIII

The iodides $C_7H_7M(CO)_2I$ (M = Mo or W) react with sodium cyclopentadienide in tetrahydrofuran solution to form orange $C_5H_5M(CO)_2C_7H_7$ (M = Mo or W) which appear to have structure **IX** (M = Mo or W) with the π-C_7H_7 ring bonded to the metal atom through only three carbon atoms rather than the usual seven carbon atoms. The iodide $C_7H_7Mo(CO)_2I$ reacts with $NaMn(CO)_5$ to form green $(CO)_5Mn$—$Mo(CO)_2C_7H_7$ which has a manganese–molybdenum bond holding the two parts of the molecule together.

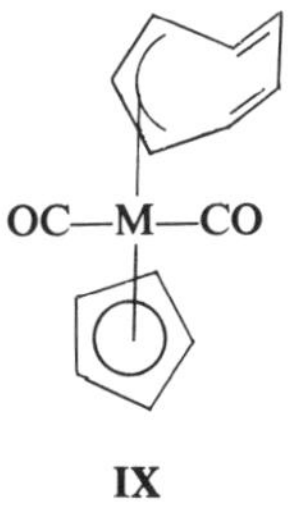

IX

The metal tricarbonyl derivatives of cyclooctatetraene also have some properties of interest. The proton NMR spectra of the $C_8H_8M(CO)_3$ (X: M = Cr, Mo, or W) compounds are temperature-dependent [29], indicative of *fluxional* properties. In the systems the metal atom travels around the C_8H_8 ring at higher temperatures but remains fixed at lower temperatures, on a time scale relative to that of the NMR measurement. Protonation of

$C_8H_8Mo(CO)_3$ gives solutions containing the $[C_8H_9Mo(CO)_3]^+$ cation, believed on the basis of NMR spectra to have the homotropylium structure **XI** [*30*]. Reaction of $C_8H_8Mo(CO)_3$ (**X**: M = Mo) with carbon monoxide for a few minutes at room temperature and atmospheric pressure gives the tetracarbonyl $C_8H_8Mo(CO)_4$ (**XII**). In this carbonylation reaction the tridentate cyclooctatetraene ligand is converted to a bidentate cyclooctatetraene ligand [*31*].

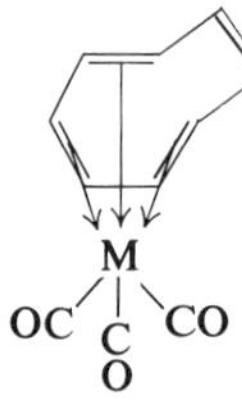

X	**XI**	**XII**

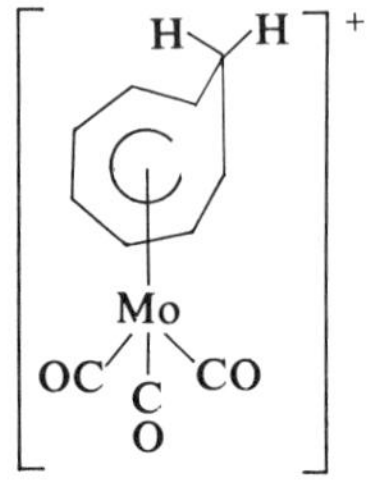

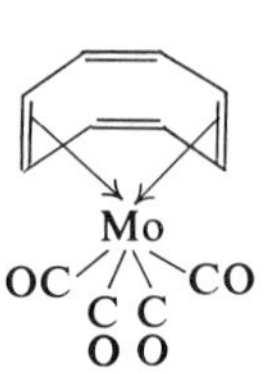

Several types of complexes have been obtained from acetylenes and carbonyl derivatives of chromium, molybdenum, and tungsten. Of particular interest are the (alkyne)$_3$MCO (M = Mo or W) complexes obtained from the alkynes, and the acetonitrile complexes $(CH_3CN)_3M(CO)_3$ (M = Mo or W). Diphenylacetylene, methyl phenyl acetylene, and hexyne-2 (diethylacetylene) undergo this reaction [*32*]. Hexafluorobutyne-2 reacts in a different manner with the acetonitrile complexes $(CH_3CN)_3M(CO)_3$, giving the white, very stable, volatile complexes $(CF_3C_2CF_3)_3MNCCH_3$ (M = Mo or W) with retention of the acetonitrile ligand rather than the carbonyl ligand [*23*]. In the (alkyne)$_3$ML complexes, the NMR spectra show the three alkyne ligands to be equivalent, but the two ends of each alkyne to be different. Structure **XIII** has been proposed for these complexes. Molecular orbital theory [*33*] can

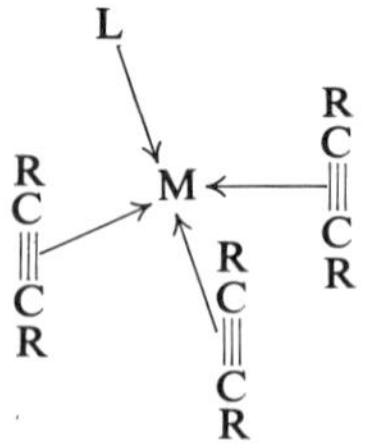

XIII

provide justification for three alkyne ligands to donate a total of ten electrons to the metal atom, giving the rare-gas configuration for the central metal atom after allowing for the two electrons donated by the ligand L.

Substituted Octahedral Metal Carbonyls

Numerous compounds can be obtained in which one or more carbonyl groups are replaced by other Lewis base ligands including tertiary phosphines, tertiary arsines, tertiary stibines, amines, pyridines, isonitriles, etc. Many such compounds can be prepared by heating the metal hexacarbonyl with the ligand in an inert solvent or in a sealed tube. Such reactions are generally carried out in the temperature range 100°–150°, but in special cases they can be carried out at temperatures as low as 50° or as high as 250°. In many cases ultraviolet irradiation may be used to effect reaction between the metal hexacarbonyl and a Lewis base ligand; this is particularly useful for thermally unstable complexes. In other cases the necessary reaction temperature may be lowered drastically (sometimes to room temperature) by substituting a halopentacarbonylmetallate $M(CO)_5I^-$ for the corresponding metal hexacarbonyl.

There are some syntheses of substituted octahedral metal carbonyls possessing a stereospecificity based on the configuration of the ligands in the starting material. Thus cycloheptatrienetricarbonylmolybdenum reacts with many Lewis bases with rapid replacement of the cycloheptatriene ring but retention of the three carbonyl groups, giving the complexes cis-$L_3M(CO)_3$. In cases of some L ligands with appreciable steric hindrance, the originally formed cis-$L_3M(CO)_3$ decomposes, giving either cis-$L_2M(CO)_4$ or possibly trans-$L_2M(CO)_4$ (e.g., L = tris(dimethylamino)phosphine $[(CH_3)_2N]_3P$). Norbornadienetetracarbonylmolybdenum similarly reacts stereospecifically with many Lewis bases, with rapid replacement of the norbornadiene ring, giving cis-$L_2M(CO)_4$. Only in cases where the two adjacent ligands in cis-$L_2M(CO)_4$ might be too crowded does isomerization to trans-$L_2M(CO)_4$ occur.

In general, the substituted octahedral metal carbonyl derivatives can be identified from their $v(CO)$ infrared frequencies. The pattern of these infrared

TABLE III-1

PREDICTED PATTERNS OF $v(CO)$ FREQUENCIES FOR SUBSTITUTED OCTAHEDRAL METAL CARBONYLS

Compound type	Point group[a]	Structure	Symmetry species of $v(CO)$	Activity[b]	Approx. intensity[c]
$M(CO)_6$	O_h		A_{1g} E_g T_{1u}	R R IR	— — Strong

TABLE III-1–*continued.*

Compound type	Point group[a]	Structure	Symmetry species of $\nu(CO)$	Activity[b]	Approx. intensity[c]
$LM(CO)_5$	C_{4v}	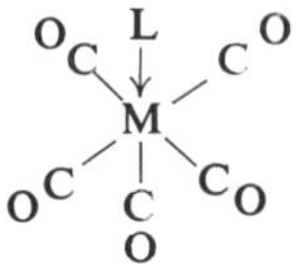	$A_1^{(1)}$ $A_1^{(2)}$ B_1 E	IR, R IR, R R IR, R	Medium Weak — Strong
trans-$L_2M(CO)_4$	D_{4h}	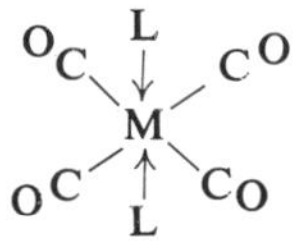	A_{1g} B_{1g} E_u	R R IR	— — Strong
cis-$L_2M(CO)_4$	C_{2v}	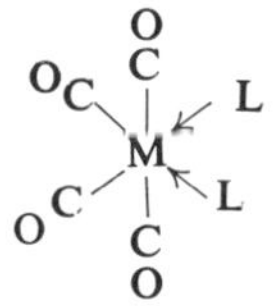	$A_1^{(1)}$ $A_1^{(?)}$ B_1 B_2	IR, R IR, R IR, R IR, R	Weak Medium Strong Medium
trans-$L_3M(CO)_3$	C_{2v}	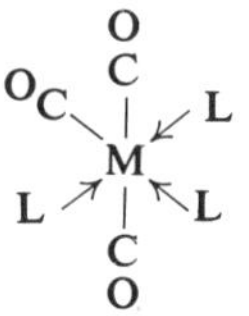	A_1 A_1 B_1	IR, R IR, R IR, R	Medium Medium Medium
cis-$L_3M(CO)_3$	C_{3v}	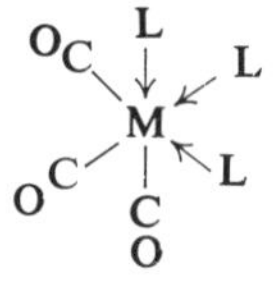	A_1 E	IR, R IR, R	Medium Strong
trans-$L_4M(CO)_2$	D_{4h}		A_{1g} A_{2u}	R IR	— Strong
cis-$L_4M(CO)_2$	C_{2v}	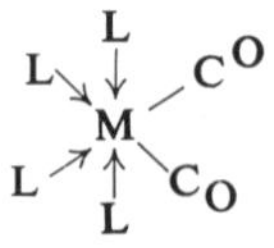	A_1 B_1	IR, R IR, R	Medium Medium

TABLE III-1—*continued.*

Compound type	Point group[a]	Structure	Symmetry species of $\nu(CO)$	Activity[b]	Approx. intensity[c]
L_5MCO	C_{4v}		A_1	IR, R	Medium

[a] The local symmetry of the ligands around the metal atom is considered.

[b] IR = infrared active; R = Raman active.

[c] The relative intensities of the infrared-active $\nu(CO)$ frequencies in the infrared spectra are considered.

frequencies can be predicted by group theoretical methods [34]. Table III-1 lists the predicted patterns of $\nu(CO)$ frequencies for the nine possible carbonyl derivatives of the type $L_{6-n}M(CO)_n$ $(1 \leqslant n \leqslant 6)$.

Cyclopentadienylmetal Carbonyl Derivatives

In an earlier section of this chapter, the reactions between various arenes and the metal hexacarbonyls to give (arene)$M(CO)_3$ (M = Cr, Mo, or W) derivatives were discussed. The cyclopentadienide anion having a similar "aromatic sextet" may be regarded as a negatively charged arene and, as such, reacts with the metal hexacarbonyls to give the $[C_5H_5M(CO)_3]^-$ anion according to the following equation [35a, 35b]:

$$M(CO)_6 + NaC_5H_5 \xrightarrow[60°-160°]{ethers} Na[C_5H_5M(CO)_3] + 3\ CO$$
$$(yellow)$$

where M = Cr, Mo, or W. The ethereal solvents suitable for this reaction range from boiling tetrahydrofuran ($\sim$65°) for conversion of $Mo(CO)_6$ to $Na[C_5H_5Mo(CO)_3]$, to boiling diglyme ($\sim$160°) for conversion of $Cr(CO)_6$ to $Na[C_5H_5Cr(CO)_3]$. Boiling dimethylformamide (153°) can also be used as a solvent for this reaction. The molybdenum and tungsten $Na[C_5H_5M(CO)_3]$ salts appear to be much easier to obtain than the analogous chromium salt and, therefore, have been studied in much greater detail.

Cyclopentadienylmetal carbonyl derivatives of molybdenum and tungsten can also be prepared by using cyclopentadiene rather than[1] sodium cyclopentadienide as the source of the π-C_5H_5 group. Thus hexacarbonylmolybdenum reacts with cyclopentadiene at temperatures above 150° to give red-violet $[C_5H_5Mo(CO)_3]_2$ [36]. This compound has structure **XIV** with the two

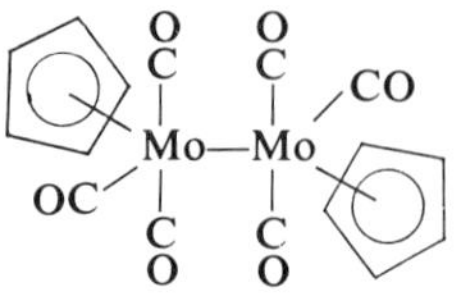

XIV

halves of the molecule held together only by a molybdenum–molybdenum bond [37]. The compound $[C_5H_5Mo(CO)_3]_2$ is of historical significance in representing the first case where a bond between two transition metals was recognized as being sufficiently strong to hold together two halves of a stable molecule. The preparation of $[C_5H_5Mo(CO)_3]_2$ is most conveniently carried out by using the dedimerization of dicyclopentadiene at its boiling point as the source of cyclopentadiene; it is merely necessary to reflux together $Mo(CO)_6$ and dicyclopentadiene at the boiling point for several hours to achieve formation of $[C_5H_5Mo(CO)_3]_2$ in reasonable yields [38]. Interestingly enough, pentamethylcyclopentadiene does *not* form the analogous $[(CH_3)_5C_5Mo(CO)_3]_2$ when heated with $Mo(CO)_6$ at temperatures above 135°. Instead, $Mo(CO)_6$ and pentamethylcyclopentadiene under these conditions form a compound $[(CH_3)_5C_5Mo(CO)_2]_2$ with two less carbonyl groups and with an assumed (not yet proven) structure **XV** with a molybdenum–molybdenum triple bond [39].

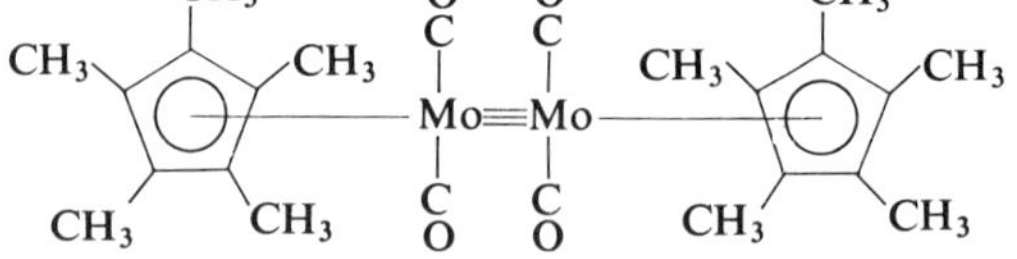

XV

The anions $[C_5H_5M(CO)_3]^-$ (M = Cr, Mo, or W) have been extremely useful intermediates for the preparation of other cyclopentadienylmetal carbonyl derivatives of these metals. The nucleophilicity of the $[C_5H_5M(CO)_3]^-$ anions increases with increasing size of the metal atom [40]; the following relative nucleophilicities were found: $[C_5H_5Cr(CO)_3]^-$ (4); $[C_5H_5Mo(CO)_3]^-$ (67); $[C_5H_5W(CO)_3]^-$ (~500). Reaction of the $[C_5H_5M(CO)_3]^-$ (M = Mo or W; occasionally also Cr) anions with a variety of inorganic and organic halides gives nonionic $RM(CO)_3C_5H_5$ derivatives according to the following equation [34]:

$$Na[C_5H_5M(CO)_3] + RX \longrightarrow RM(CO)_3C_5H_5 + NaX$$

This synthetic technique is useful for the preparation of a variety of compounds with molybdenum or tungsten σ-bonded to carbon, germanium, tin, lead, mercury, gold, and other transition metals. Halides reacting according to the above equation include methyl iodide, ethyl bromide, allyl chloride, benzyl chloride, chloromethyl methyl sulfide, heptafluorobutyryl chloride, trimethyltin chloride, triphenyllead chloride, $(C_6H_5)_3PAuCl$, and $C_5H_5Fe(CO)_2I$. The $RM(CO)_3C_5H_5$ compounds where R is bonded to the molybdenum or tungsten atom through carbon, germanium, tin, or lead are pale yellow solids which, in most cases, are sufficiently volatile to sublime in vacuum. The tungsten derivatives are more stable than the molybdenum derivatives, both to thermal decomposition (often giving $[C_5H_5M(CO)_3]_2$) and air oxidation. In most cases, the chromium derivatives $RCr(CO)_3C_5H_5$ appear to be too unstable for isolation.

A further characteristic reaction of the anions $[C_5H_5M(CO)_3]^-$ is their reaction with nonoxidizing acids to give the isolable yellow "hydrides" $HM(CO)_3C_5H_5$ (M = Cr, Mo, or W) (**XVI**) [35a, 35b]. Since these hydrides are but weakly acidic (pK ~7), even relatively weak acids such as acetic acid will suffice for their liberation. The chromium and molybdenum $HM(CO)_3C_5H_5$ derivatives (**XVI**: M = Cr or Mo) are rather unstable since they oxidize

rapidly in air and liberate hydrogen above ~50°, giving the binuclear derivatives $[C_5H_5M(CO)_3]_2$ in both types of reactions. The tungsten derivative $HW(CO)_3C_5H_5$ is much more stable since it liberates hydrogen only above ~180° and can be handled in air for several minutes without appreciable decomposition; however, upon exposure to light, the pale yellow $HW(CO)_3C_5H_5$ soon turns pink, apparently because of some tungsten–hydrogen bond cleavage forming $[C_5H_5W(CO)_3]_2$. The hydrides $HM(CO)_3C_5H_5$ (M = Cr, Mo, or W) are useful synthetic intermediates. For many synthetic applications, they can be generated in tetrahydrofuran solution by mixing $Na[C_5H_5M(CO)_3]$ and acetic acid, and then used without isolation.

The sodium salts $Na[C_5H_5M(CO)_3]$ (M = Mo or W) are useful for the preparation of unusual organometallic derivatives such as the following.

π-ALLYL DERIVATIVES [41]

$$NaM(CO)_3C_5H_5 + CH_2{=}CHCH_2Cl \xrightarrow{THF} \sigma\text{-}C_3H_5M(CO)_3C_5H_5 + NaCl$$

(yellow liquid,
air-sensitive)

$$\sigma\text{-}C_3H_5M(CO)_3C_5H_5 \xrightarrow[25°]{UV} (\pi\text{-}C_3H_5)M(CO)_2(\pi\text{-}C_5H_5) + CO$$

[volatile yellow crystals
(**XVII**) obtained in
~50% yield]

π-BENZYL DERIVATIVE [42]

$$NaMo(CO)_3C_5H_5 + C_6H_5CH_2Cl \xrightarrow{THF} \sigma\text{-}C_6H_5CH_2Mo(CO)_3C_5H_5 + NaCl$$

(yellow crystals)

$$\sigma\text{-}C_6H_5CH_2Mo(CO)_3C_5H_5 \xrightarrow[25°]{UV} (\pi\text{-}C_6H_5CH_2)Mo(CO)_2(\pi\text{-}C_5H_5) + CO$$

(volatile red crystals obtained in
only ~5% yield)

The proton NMR spectrum of the π-benzyl derivative at $-40°$ is consistent with the "fixed" π-benzyl structure **XVIII**. However, upon warming the solutions of the π-benzyl derivative, the various resonances in the NMR spectrum coalesce in a manner consistent with that expected if the two sides of the π-benzyl ligand are becoming equivalent through a fluxional process. The π-benzyl derivative $(\pi\text{-}C_6H_5CH_2)Mo(CO)_2(\pi\text{-}C_5H_5)$ is thus an example of a fluxional molecule.

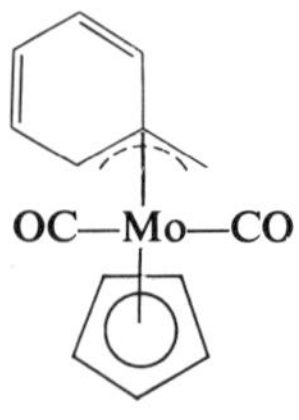

OC—Mo—CO

XVIII

π-CH_3SCH_2 DERIVATIVES [43]

$$NaM(CO)_3C_5H_5 + CH_3SCH_2Cl \longrightarrow CH_3SCH_2M(CO)_3C_5H_5 + NaCl$$

(yellow crystals)

$$CH_3SCH_2M(CO)_3C_5H_5 \xrightarrow[heat]{UV\ or} (\pi\text{-}CH_2SCH_2)Mo(CO)_2C_5H_5 + CO$$

(yellow crystals)

The structure **XIX** for $(\pi\text{-}CH_3SCH_2)Mo(CO)_2C_5H_5$ is similar to that of a substituted π-allyl derivative, but with an —S— atom replacing a —CH=CH—

group in the π-bonded ligand as in the relationship between thiophene and benzene.

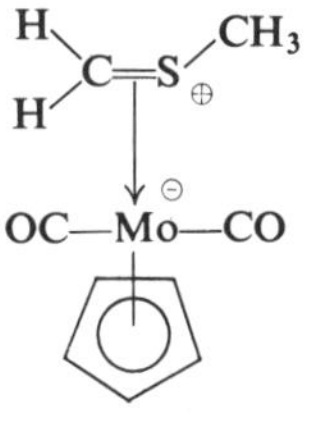

$$\begin{array}{c} H \\ \diagdown \\ H \diagup \end{array} C \!\!=\!\! S \underset{\oplus}{\diagup} CH_3$$

OC—Mo—CO

XIX

ARYLAZO DERIVATIVES [44]

$$NaMo(CO)_3C_5H_5 + RN_2{}^+BF_4{}^- \xrightarrow[-78°]{THF} RN_2Mo(CO)_2C_5H_5 + CO + NaBF_4$$

[red crystals (**XX**)]

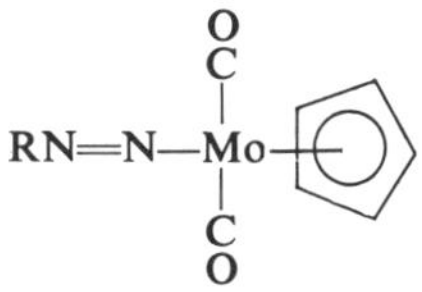

$$RN\!\!=\!\!N\!\!-\!\!Mo \overset{\overset{O}{\overset{\|}{C}}}{\underset{\underset{O}{\underset{\|}{C}}}{}}$$

XX

where R = phenyl, *p*-methoxyphenyl, *p*-tolyl, and *p*-nitrophenyl. The *p*-tolyl derivative reacts with triphenylphosphine in boiling methylcyclohexane ($\sim 100°$) to give the orange substitution product

$$(p\text{-}CH_3C_6H_4N_2)Mo(CO)[P(C_6H_5)_3](C_5H_5)$$

with four different ligands bonded to the central molybdenum atom.

Cyclopentadienylmetal carbonyl halides of molybdenum and tungsten (not chromium) of the types $C_5H_5M(CO)_3X$ and $C_5H_5Mo(CO)_2X_3$ (both with favored eighteen-electron rare-gas electronic configurations) have been prepared by reactions such as the following:

$$C_5H_5M(CO)_3H + CCl_4 \longrightarrow C_5H_5M(CO)_3Cl + CHCl_3 \quad [35a] \quad \text{(a)}$$
$$\text{(M = Mo or W)} \qquad\qquad \text{(red)}$$

$$C_5H_5M(CO)_3H + BrNC_4H_4O_2 \longrightarrow C_5H_5M(CO)_3Br + HNC_4H_4O_2 \quad [35a] \quad \text{(b)}$$
$$\text{(M = Mo or W; BrNC}_4\text{H}_4\text{O}_2 = \qquad\qquad \text{(red)}$$
$$\text{N-bromosuccinimide)}$$

$$[C_5H_5M(CO)_3]_2 + I_2 \longrightarrow 2\,C_5H_5M(CO)_3I \quad [45] \qquad\qquad \text{(c)}$$
$$\text{(M = Mo or W)} \qquad\qquad \text{(red-black, mp 134°)}$$

$$[C_5H_5Mo(CO)_3]_2 + 3\,X_2 \xrightarrow{\text{benzene}} 2\,C_5H_5Mo(CO)_2X_3 + 2\,CO \quad [46] \quad \text{(d)}$$
$$\text{(X = Br or I)} \qquad\qquad \text{(brown)}$$

The $C_5H_5M(CO)_3X$ halides react with various ligands in one of the following two ways. (a) Replacement of carbonyl groups, with the ligand introduced forming a nonionic derivative according to the following scheme:

$$C_5H_5M(CO)_3X + L \longrightarrow C_5H_5M(CO)_2LX + CO$$

This type of reaction occurs particularly readily with tricovalent phosphorus ligands such as $[(CH_3)_2N]_3P$ and $(C_6H_5)_3P$. (b) Replacement of the halogen with the ligand introduced without carbon monoxide evolution forming an ionic derivative according to the following scheme [47]:

$$C_5H_5M(CO)_3X + L \longrightarrow [C_5H_5M(CO)_3L]X$$

Addition of anhydrous aluminum halides to the mixture of $C_5H_5M(CO)_3X$ and the ligand increases the probability that the reaction will proceed according to scheme (b) rather than according to scheme (a). The complex cations formed in this reaction can often be conveniently isolated as hexafluorophosphates by hydrolysis of the reaction mixture followed by addition of ammonium hexafluorophosphate.

The yellow cyclopentadienylmetal tetracarbonyl cations $[C_5H_5M(CO)_4]^+$ (M = Mo or W) can be synthesized by reaction of a mixture of $C_5H_5M(CO)_3Cl$ and aluminum chloride with carbon monoxide under pressure [48]. Similarly, the red benzenecyclopentadienylmetal carbonyl cations $[C_5H_5M(CO)C_6H_6]^+$ (XXI: M = Mo or W) can be synthesized by addition of $C_5H_5M(CO)_3Cl$ to boiling benzene containing aluminum chloride [49].

Reactions of reactive organosulfur compounds with cyclopentadienylmetal carbonyl derivatives of chromium, molybdenum, and tungsten can often be used to prepare unusual complexes containing both π-cyclopentadienyl and organosulfur ligands. The following reactions of dimethyl disulfide and bis-(trifluoromethyl)dithietene $(CF_3)_2C_2S_2$ illustrate some of the unusual types of complexes which can be prepared:

$$[C_5H_5Mo(CO)_3]_2 + 2\ CH_3SSCH_3 \xrightarrow{135°} [C_5H_5Mo(SCH_3)_2]_2 + 6\ CO \quad [50] \quad (a)$$
$$\text{(brown, XXII)}$$

$$2\ C_5H_5Mo(CO)_3H + 2\ CH_3SSCH_3 \xrightarrow{25°} [C_5H_5Mo(CO)_2SCH_3]_2 + 2\ CH_3SH$$
$$\text{(red-black, XXIII)} \qquad + 2\ CO \quad [51]$$
$$(b)$$

$$2\ C_5H_5Mo(CO)_2NO + 2\ (CF_3)_2C_2S_2 \longrightarrow [C_5H_5Mo(CO)S_2C_2(CF_3)_2]_2 + 4\ CO \quad [52]$$
$$\text{(brown, XXIV)} \qquad (c)$$

$$C_5H_5W(CO)_2NO + 2\ (CF_3)_2C_2S_2 \longrightarrow C_5H_5W[S_2C_2(CF_3)_2]_2 + 2\ CO + NO \quad [52]$$
$$\text{[dark green, XXV,} \qquad (d)$$
$$\text{paramagnetic}$$
$$\text{(1.68 BM)]}$$

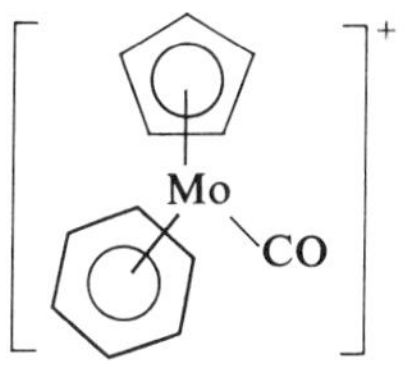

XXI

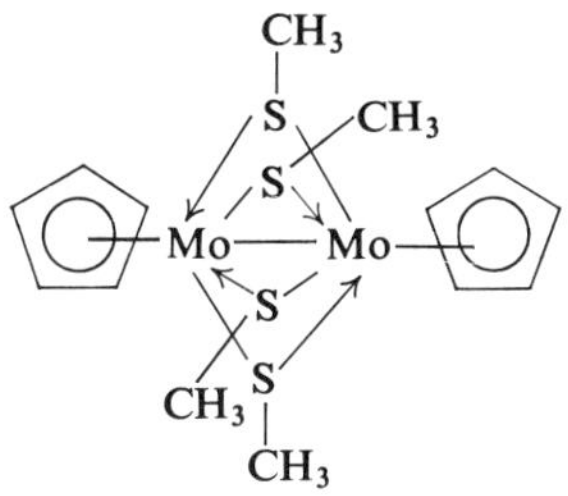

XXII

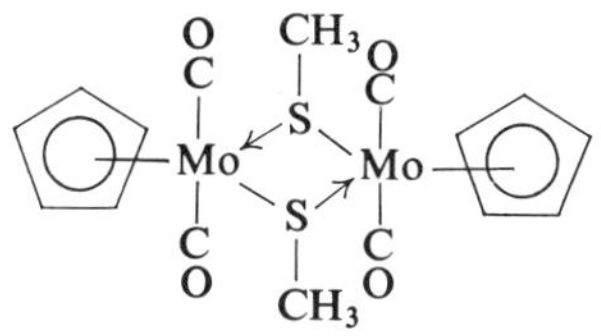

XXIII

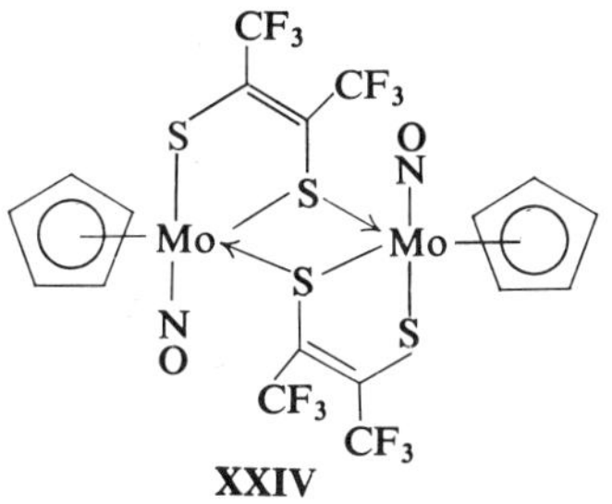

XXIV

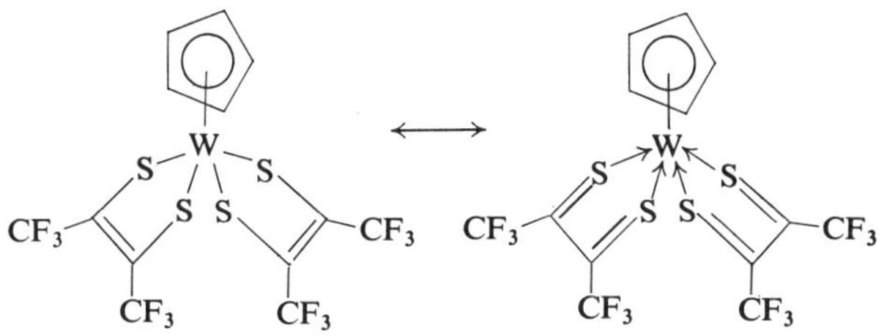

XXV

The compound [C$_5$H$_5$Mo(SCH$_3$)$_2$]$_2$ (**XXII**) is a rare example of a complex containing four bridging groups.

Cyclopentadienylmetal Nitrosyl Derivatives

A variety of cyclopentadienylmetal nitrosyl derivatives of chromium, molybdenum, and tungsten can be prepared. These include the cyclopenta-dienylmetal dicarbonyl nitrosyls, C$_5$H$_5$M(CO)$_2$NO (**XXVI**: M = Cr, Mo, or W), volatile orange solids which are best prepared by the following reactions:

$$[C_5H_5Cr(CO)_3]_2 + 2\ NO \xrightarrow[25°]{benzene} 2\ C_5H_5Cr(CO)_2NO + 2\ CO \quad [53] \quad (a)$$

$$C_5H_5M(CO)_3H + p\text{-}CH_3C_6H_4SO_2N(NO)CH_3 \longrightarrow$$

$$C_5H_5M(CO)_2NO + \text{other products} \quad [35a] \quad (b)$$

Of the three $[C_5H_5M(CO)_3]_2$ compounds only the chromium derivative reacts with nitric oxide to form $C_5H_5Cr(CO)_2NO$ as in Eq. (a) above. Apparently the metal–metal bonds in the molybdenum and tungsten $[C_5H_5M(CO)_3]_2$ (M = Mo or W) are too stable to be cleaved by nitric oxide.

The molybdenum derivative $C_5H_5Mo(CO)_2NO$ can be converted to other cyclopentadienylmolybdenum nitrosyl derivatives by reactions such as the following:

$$2\ C_5H_5Mo(CO)_2NO + 2\ I_2 \xrightarrow{CH_2Cl_2} [C_5H_5Mo(NO)I_2]_2 + 4\ CO \quad [54] \quad (a)$$
$$\text{(brown-purple precipitate)}$$

$$[C_5H_5Mo(NO)I_2]_2 + 2\ L \xrightarrow{CH_2Cl_2} 2\ C_5H_5Mo(NO)I_2L \quad [54] \quad (b)$$
$$\text{(red)}$$

$$(L = (C_6H_5)_3P,\ (C_6H_5O)_3P,\ \text{or pyridine})$$

$$[C_5H_5Mo(NO)I_2]_2 + 2\ TlC_5H_5 \xrightarrow{THF} 2\ (C_5H_5)_2Mo(NO)I + 2\ TlI \quad [55] \quad (c)$$
$$\text{(green-brown)}$$

$$[C_5H_5Mo(NO)I_2]_2 + 4\ TlC_5H_5 \xrightarrow{THF} 2\ (C_5H_5)_3MoNO + 4\ TlI \quad [55a] \quad (d)$$

Sodium cyclopentadienide does not react in this manner because it is too reactive.

$$2\ (C_5H_5)_2Mo(NO)I + 2\ CH_3MgBr \xrightarrow{ether} \xrightarrow{H_2O} 2\ (C_5H_5)_2Mo(NO)CH_3 +$$
$$\text{(brown, volatile at } 70°/0.1 \text{ mm)}$$
$$MgBr_2 + MgI_2 \quad [55] \quad (e)$$

The compounds $(C_5H_5)_2Mo(NO)Y$ [Y = I, reaction (c); Y = σ-C_5H_5, reaction (d); Y = CH_3, reaction (e)] are particularly unusual, since in order to have the favored eighteen-electron rare-gas electronic configuration, one of the two π-cyclopentadienyl rings must be bonded to the molybdenum atom through only three, rather than the usual five, carbon atoms, as indicated in structure **XXVII** (Y = I, σ-C_5H_5, or CH_3).

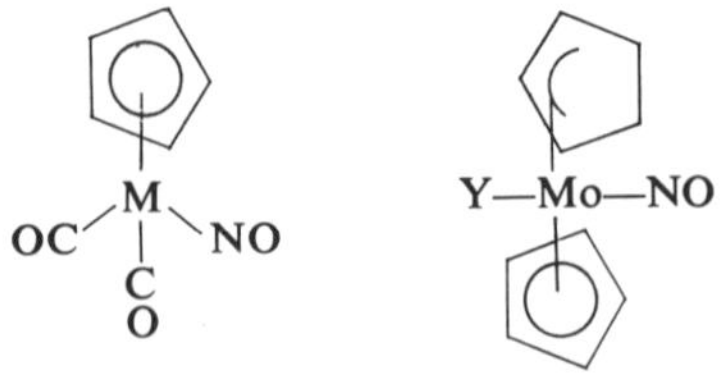

XXVI   **XXVII**

Most of the cyclopentadienylmetal nitrosyl derivatives discussed up to this point have been molybdenum compounds prepared from $C_5H_5Mo(CO)_2NO$.

A variety of cyclopentadienylchromium nitrosyl derivatives are also known. These are generally prepared from $C_5H_5Cr(NO)_2Cl$, a stable olive-green solid which can be obtained by successive treatment of anhydrous chromium trichloride with sodium cyclopentadienide and nitric oxide [56, 57]. The molybdenum analogue $C_5H_5Mo(NO)_2Cl$ is also known, having been prepared by the following sequence of reactions:

$$Mo(CO)_6 + 2\ ClNO \xrightarrow{CH_2Cl_2} \tfrac{1}{n}\ [Mo(NO)_2Cl_2] + 6\ CO \quad [58] \quad (a)$$

$$\tfrac{1}{n}[Mo(NO)_2Cl_2]_n + TlC_5H_5 \xrightarrow{THF} C_5H_5Mo(NO)_2Cl + TlCl \quad [55] \quad (b)$$

The $C_5H_5Mo(NO)_2Cl$ is a yellow-green crystalline solid which decomposes unpredictably upon standing at room temperature. It has not yet been used for the preparation of other cyclopentadienylmolybdenum nitrosyl derivatives.

The chromium compound $C_5H_5Cr(NO)_2Cl$ undergoes several reactions of interest. Treatment of $C_5H_5Cr(NO)_2Cl$ with alkylmagnesium halides gives the volatile green σ-alkyl derivatives $RCr(NO)_2C_5H_5$ ($R = CH_3$, C_2H_5, or σ-C_5H_5) [35a]. Reduction of $C_5H_5Cr(NO)_2Cl$ with aqueous sodium borohydride gives the red-violet bimetallic $[C_5H_5Cr(NO)_2]_2$ with structure **XXVIII**, containing both terminal [$\nu(NO) = 1672$ cm^{-1}] and bridging [$\nu(NO) = 1505$ cm^{-1}] nitrosyl groups on the basis of its infrared spectrum [59]. Reaction of $C_5H_5Cr(NO)_2Cl$ with a variety of Lewis bases (amines, phosphines, isonitriles,

XXVIII

etc.) gives deep green monometallic derivatives of the type $C_5H_5Cr(NO)(L)Cl$ [60]. These substitution products have only a seventeen-electron configuration and, therefore, exhibit the expected paramagnetism for one unpaired electron. Reaction of $C_5H_5Cr(NO)_2Cl$ with carbon monoxide under pressure in the presence of aluminum chloride gives the cation $[C_5H_5Cr(NO)_2CO]^+$, isolated as its green-brown hexafluorophosphate [57]. This formation of $[C_5H_5Cr(NO)_2CO]^+$ from $C_5H_5Cr(NO)_2Cl$ resembles the formation of $[C_5H_5M(CO)_4]^+$ ($M = Mo$ or W) from $C_5H_5M(CO)_3Cl$ discussed above.

Biscyclopentadienyl Derivatives of Chromium, Molybdenum, and Tungsten

Several biscyclopentadienyl derivatives of chromium are known. Reaction of anhydrous chromium(III) chloride with sodium cyclopentadienide in tetrahydrofuran solution gives dark red, air-sensitive, volatile biscyclopenta-dienylchromium (chromocene) $(C_5H_5)_2Cr$, mp 173° [61]. This chromium complex $(C_5H_5)_2Cr$ has a sixteen-electron configuration, which is two less than the rare-gas configuration. These two "holes" are unpaired, resulting in two unpaired electrons in accord with the observed paramagnetism. Excess oxygen instantly destroys $(C_5H_5)_2Cr$; however, a deficiency of oxygen converts $(C_5H_5)_2Cr$ into the blue tetrametallic oxide $[C_5H_5CrO]_4$ which is decomposed by excess oxygen [62]. Certain organic halides oxidize $(C_5H_5)_2Cr$ to the air-sensitive $[(C_5H_5)_2Cr]^+$ cation, e.g.:

$$2\ (C_5H_5)_2Cr + 2\ CH_2{=}CHCH_2I \longrightarrow 2\ [(C_5H_5)_2Cr]I + C_6H_{10} \qquad [63] \qquad (a)$$

$$(C_5H_5)_2Cr + CCl_4 \longrightarrow [(C_5H_5)_2Cr][C_5H_5CrCl_3] + \text{other products} \qquad [64] \\ (b)$$

Biscyclopentadienyl derivatives of molybdenum and tungsten of the type $(C_5H_5)_2M$ have not yet been prepared. Reaction of $MoCl_5$ or WCl_6 with sodium cyclopentadienide gives a low yield of the dihydrides $(C_5H_5)_2MH_2$ (**XXIX**: M — Mo or W); the yield is greatly improved by adding sodium borohydride as a source of the two hydrogen atoms directly bonded to the metal atom [65].

The dihydrides $(C_5H_5)_2MH_2$ (**XXIX**: M = Mo or W) are volatile, air-sensitive, yellow solids which are soluble in organic solvents. These dihydrides do not dissolve in water, but dissolve in aqueous solutions of strong acids to give solutions containing the $[(C_5H_5)_2MH_3]^+$ cations; solid salts of these cations have been isolated. The basic character of $(C_5H_5)_2WH_2$ is also demonstrated by its ability to form an adduct $(C_5H_5)_2WH_2 \cdot BF_3$ upon reaction with boron trifluoride [66] and the adducts $(C_5H_5)_2WH_2 \cdot M(CO)_5$ (M=Cr, Mo, or W) upon reaction with $C_4H_8OM(CO)_5$ [66a]. The dihydrides $(C_5H_5)_2MH_2$ can be converted to the corresponding dihalides $(C_5H_5)_2MX_2$ (M = Mo or W; X = Cl, Br, or I) with certain halogenating agents, e.g., CCl_4, $CHBr_3$, and I_2 [67]. Stronger halogenating agents (e.g., Cl_2 and Br_2) oxidize the initially formed neutral dihalide $(C_5H_5)_2MX_2$ into the corresponding cation $[(C_5H_5)_2MX_2]^+$, often obtained as an HX_2^- salt.

H—M—H

XXIX

Other Chromium, Molybdenum, and Tungsten Derivatives with Two π-Bonded Rings

The compound $(C_5H_5)_2Cr$ (two five-membered rings; sixteen-electron configuration) has already been discussed. Other types of compounds with two π-bonded rings include $(C_6H_6)_2M$ (two six-membered rings; eighteen-electron configuration), $C_5H_5MC_7H_7$ (M = Cr or Mo; one five-membered ring and one seven-membered ring, eighteen-electron configuration), and $C_5H_5MC_6H_6$ (M = Cr or Mo; one five-membered ring and one six-membered ring; seventeen-electron configuration). The complexes of the types $(C_6H_6)_2M$ and $C_5H_5MC_7H_7$ with eighteen-electron configurations can be oxidized reversibly to the corresponding cations $[(C_6H_6)_2M]^+$ and $[C_5H_5MC_7H_7]^+$ with only seventeen-electron configurations. The complexes with two rings of unequal size bonded to the metal atom are extremely difficult to prepare, especially in quantities larger than one gram.

DIBENZENE–METAL DERIVATIVES

Dibenzene–metal derivatives are best prepared by the so-called "reducing Friedel–Crafts reaction" which involves heating the anhydrous metal halide with benzene, aluminum chloride, and aluminum powder, followed by hydrolysis and further reduction in basic solution of the intermediate cation if the neutral derivative is wanted. Thus, heating a mixture of chromium(III) chloride, aluminum chloride, aluminum powder, and benzene for several hours, followed by hydrolysis and further reduction with sodium dithionite $(Na_2S_2O_4)$ in strongly alkaline solution gives brown-black, volatile dibenzene–chromium $(C_6H_6)_2Cr$ (**XXX**: M = Cr), mp 284°–285° [68]. Alkylated benzene

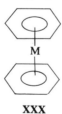

XXX

derivatives will also undergo this reaction, giving substituted dibenzene–chromium derivatives. However, benzene derivatives containing substituents with lone electron pairs (e.g., chlorobenzene, aniline, phenol, methyl benzoate, etc.) will not undergo this reaction, since they deactivate the aluminum chloride catalyst by complex formation. A similar "reducing Friedel–Crafts reaction" can be used to prepare green dibenzene–molybdenum and dibenzene–tungsten $(C_6H_6)_2M$ (**XXX**: M = Mo or W) [69]. However, in the cases of the molybdenum and tungsten derivatives, conversion of the $[(C_6H_6)_2M]^+$ cations to the neutral $(C_6H_6)_2M$ compounds can be accomplished with strong aqueous alkali alone, without addition of a reducing agent such as sodium dithionite. The dibenzene–metal derivatives of molybdenum and tungsten are much less stable to air-oxidation and thermal decomposition than the chromium compound. Furthermore, the yield of $(C_6H_6)_2W$ from tungsten hexachloride by the reducing Friedel–Crafts reaction is only ~2%.

Some chemistry of dibenzene–chromium has been investigated. It is impossible to effect electrophilic substitution of the benzene rings in dibenzene–chromium; oxidation to $[(C_6H_6)_2Cr]^+$ occurs instead. Dibenzene–chromium can be metallated with n-amylsodium, a very powerful metallating agent, according to the following equation [70]:

$$(C_6H_6)_2Cr + NaC_5H_{11} \longrightarrow NaC_6H_5CrC_6H_6 + C_5H_{12}$$

This sodium derivative reacts readily with organic carbonyl compounds RCOR' (R and R' = H, CH_3, C_6H_5, etc.) to give the alcohols $C_6H_6CrC_6H_5CRR'(OH)$ (XXXI), which, like the parent dibenzene–chromium, are brown, easily oxidized solids. Oxidations of the alcohols XXXI (R' = H) with aluminum isopropoxide in a mixture of acetone and benzene give the corresponding red ketones $C_6H_6CrC_6H_5COR$ (XXXII), which are more stable to air-oxidation but less stable to thermal decomposition than the parent alcohols.

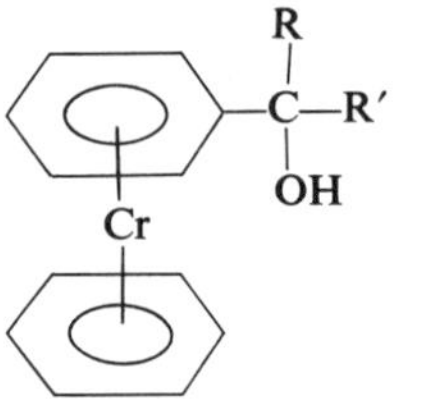

XXXI 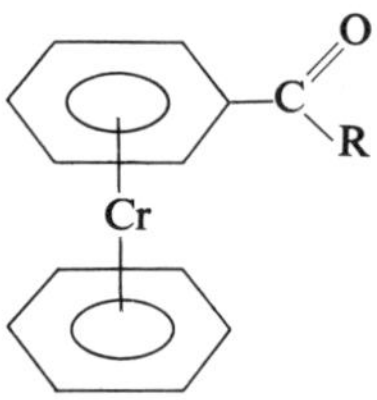   XXXII

Other reactions of dibenzene–chromium of some importance include the following: (a) facile oxidation to the $[(C_6H_6)_2Cr]^+$ cation by mild oxidizing agents such as iodine and oxygen; (b) complete replacement of both benzene ligands in dibenzene–chromium by reaction with other ligands under vigorous conditions; e.g. [71]

$$(C_6H_6)_2Cr + 6\ PF_3 \longrightarrow Cr(PF_3)_6 + 2\ C_6H_6$$

$$(C_6H_6)_2Cr + 2\ C_{15}H_{11}N_3 \longrightarrow (C_{15}H_{11}N_3)_2Cr + 2\ C_6H_6$$

To date it has not proved possible to selectively replace only one of the two benzene rings in dibenzene–chromium by heating with such ligands.

CYCLOPENTADIENYLBENZENEMETAL DERIVATIVES

The orange chromium compound $C_5H_5CrC_6H_6$ (XXXIII: M = Cr) can be prepared in about 20% yield by reaction of chromium(III) chloride with a mixture of sodium cyclopentadienide and phenylmagnesium bromide followed by hydrolysis [72]. The red molybdenum compound $C_5H_5MoC_6H_6$ (**XXXIII**: M = Mo) can be prepared by lithium aluminum hydride reduction of the salt $[C_5H_5Mo(CO)C_6H_6][PF_6]$ in tetrahydrofuran solution [73]. However, analogous lithium aluminum hydride reduction of the tungsten salt $[C_5H_5W(CO)C_6H_6][PF_6]$ did not give $C_5H_5WC_6H_6$ but, instead, gave the yellow π-cyclohexadiene tungsten hydride $C_5H_5WC_6H_8(CO)H$ (**XXXIV**).

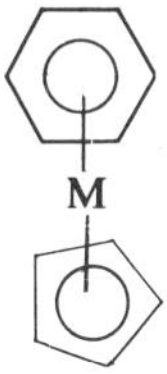

XXXIII 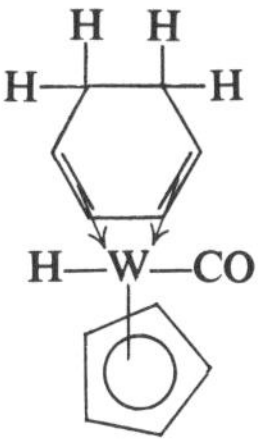   **XXXIV**

Cyclopentadienylcycloheptatrienylmetal Derivatives

The following reactions summarize methods which have been used for the preparation of green $C_5H_5CrC_7H_7$ (**XXXV**: M = Cr) and its substitution products:

$$CrCl_3 + C_5H_6 + C_7H_8 + (CH_3)_2CHMgBr \xrightarrow{\text{ether}}$$

$$C_5H_5CrC_7H_7 + \text{other products} \quad [74] \quad (a)$$
$$\text{(green, diamagnetic)}$$

$$C_5H_5CrCl_2 \cdot OC_4H_8 + C_7H_8 + (CH_3)_2CHMgBr \xrightarrow{\text{ether}} C_5H_5CrC_7H_8 \quad [75] \quad (b)$$
$$\text{(brown, paramagnetic, \textbf{XXXVI})}$$

$$2\,C_5H_5CrC_7H_8 \xrightarrow{\text{Pd/C}} 2\,C_5H_5CrC_7H_7 + H_2$$

$$C_5H_5CrC_6H_6 + C_7H_7^+ \longrightarrow$$

$$[C_5H_5CrC_7H_7]^+ + C_6H_6 \quad [76] \quad (c)$$
$$\text{(yellow, paramagnetic,}$$
$$\text{isolated as } PF_6^- \text{ salt)}$$

$$[C_5H_5CrC_7H_7]^+ \xrightarrow[\text{Na}_2\text{S}_2\text{O}_4]{\text{alkaline}} C_5H_5CrC_7H_7$$

$$C_5H_5CrC_6H_6 + RCOCl \xrightarrow[\text{CS}_2]{\text{AlCl}_3} [C_5H_5CrC_7H_6R]^+ +$$
$$\text{other products} \quad [77]$$
$$(R = CH_3 \text{ or } C_6H_5) \qquad\qquad (d)$$

$$[C_5H_5CrC_7H_6R]^+ \xrightarrow[\text{Na}_2\text{S}_2\text{O}_4]{\text{alkaline}} C_5H_5CrC_7H_6R$$

By substituting $MoCl_5$ for the $CrCl_3$ in reaction (a) above the brown air-sensitive molybdenum analogue $C_5H_5MoC_7H_7$ (**XXXV**: M = Mo) can be prepared, but only in very low yield. The dehydrogenation of the seventeen-electron complex $C_5H_5CrC_7H_8$ (**XXXVI**) to give the eighteen-electron complex $C_5H_5CrC_7H_7$ (**XXXV**: M = Cr) is an unusual illustration of the stability of the favored eighteen-electron rare-gas configuration. Reaction (d) above is an unusual example of the expansion of a π-benzene ring into a π-cycloheptatrienyl ring.

XXXV   **XXXVI**

Miscellaneous Organochromium Compounds

A few π-allyl derivatives of chromium have been prepared [78]. Thus, reaction of chromium(III) chloride with allylmagnesium bromide in diethyl ether solution at $-18°$ gives deep red triallylchromium(III) $(C_3H_5)_3Cr$, mp 77–79°, sublimes at 1°/1 mm. The formulation as a chromium(III) complex is supported by a magnetic moment of 3.78 BM, corresponding to the expected three unpaired electrons. Heating $(C_3H_5)_3Cr$ to slightly above room temperature removes one allyl group with concurrent dimerization, giving red-brown tetraallyldichromium(II) $[(C_3H_5)_2Cr]_2$, decomposes at 121°–129°, sublimes at 50°/1 mm.

Some octahedral σ-phenylchromium(III) derivatives have been prepared. Reaction of phenylmagnesium bromide with anhydrous chromium(III) chloride [or its tris(tetrahydrofuranate)] in tetrahydrofuran solution at $-20°$ gives blood red, very air-sensitive $(C_6H_5)_3Cr(OC_4H_8)_3$ [79]. Similarly, the reaction of anhydrous chromium(III) chloride with excess phenyllithium in diethyl ether solution gives the red-yellow lithium salt $Li_3[Cr(C_6H_5)_6] \cdot 2.5$ $(C_2H_5)_2O$ [80]. The octahedral coordination is necessary for the stability of these phenylchromium derivatives. Removal of the coordinated tetrahydrofuran from $(C_6H_5)_3Cr(OC_4H_8)_3$ in an attempt to prepare triphenylchromium results, instead, in the collapse of the σ-phenyl groups to give π-benzene and π-biphenyl derivatives, including dibenzene–chromium, bis(biphenyl)-chromium, and the mixed compound benzene–biphenyl–chromium. The same result can be achieved by carrying out the reaction between phenylmagnesium bromide and chromium(III) chloride in diethyl ether rather than tetrahydrofuran. Diethyl ether, unlike tetrahydrofuran, is too weak a base to stabilize a triphenylchromium fragment by forming an adduct. Thus, in diethyl ether solution the σ-phenylchromium intermediates cannot be isolated but, instead, are immediately converted to π-benzene and π-biphenyl derivatives. This was the basis for Hein's classic synthesis [81] of the "polyphenyl chromium compounds" which were much later [82] shown to be bis-π-arene chromium complexes. The lithium salt $Li_3[Cr(C_6H_5)_6] \cdot 2.5 \, (C_2H_5)_2O$ is also of interest in having the unusual property of reacting with molecular hydrogen at room temperature and atmospheric pressure in the absence of a catalyst to remove one phenyl group as benzene, giving the bright red hydride $Li_3[CrH(C_6H_5)_5] \cdot 3(C_2H_5)_2O$.

Some organochromium compounds have been prepared which also contain coordinated water ligands [83, 84]. Reaction of an aqueous solution of chromium(II) with benzyl chloride gives a mixture of green $[Cr(H_2O)_5Cl]^{2+}$, violet $[Cr(H_2O)_6]^{3+}$, and yellow-orange $[C_6H_5CH_2Cr(H_2O)_5]^{2+}$. This benzyl-pentaquochromium(III) cation has a half-life of $\sim 1\frac{1}{2}$ days in aqueous solution. Similarly, reaction of an aqueous solution of chromium(II) with chloroform gives a mixture of green $[Cr(H_2O)_5Cl]^{2+}$ and red $[Cl_2CHCr(H_2O)_5]^{2+}$.

REFERENCES

1. A. Job and A. Cassal, *Compt. Rend.* **183**, 392 (1926).
2. G. Natta, R. Ercoli, F. Calderazzo, and A. Rabizzoni, *J. Am. Chem. Soc.* **76**, 3611 (1957).
3. E. O. Fischer, W. Hafner, and K. Öfele, *Ber.* **92**, 3050 (1959).
4. K. A. Kocheskov, A. N. Nesmeyanov, M. M. Nadj, and I. M. Rossinskaya, *Compt. Rend. Acad. Sci. URSS* **26**, 54 (1940).
5. A. N. Nesmeyanov, E. P. Mikheev, K. N. Anisimov, V. L. Volkov, and Z. P. Valueva, *Zh. Neorgan. Khim.* **4**, 249 and 503 (1959).
6. E. O. Fischer and K. Öfele, *Ber.* **93**, 1156 (1960).
7. E. W. Abel, I. S. Butler, and J. G. Reid, *J. Chem. Soc.* p. 2068 (1963).
8. H. Behrens and H. Zizlsperger, *Z. Naturforsch.* **16b**, 349 (1961).
9. R. B. King, *Inorg. Chem.* **3**, 1039 (1964); M. C. Ganorkar and M. H. B. Stiddard, *J. Chem. Soc.* p. 3494 (1965).
10. H. D. Murdoch and R. Henzi, *J. Organometal. Chem.* (*Amsterdam*) **5**, 166 and 463 (1966).
11. H. D. Murdoch, *J. Organometal. Chem.* (*Amsterdam*) **4**, 119 (1965).
12. R. Colton and I. B. Tomkins, *Australian J. Chem.* **19**, 1143 and 1519 (1966).
13. H. Behrens and R. Weber, *Z. Anorg. Allgem. Chem.* **291**, 123 (1957).
14. R. G. Hayter, *J. Am. Chem. Soc.* **88**, 4376 (1966).
15. L. B. Handy, P. M. Treichel, and L. F. Dahl, *J. Am. Chem. Soc.* **88**, 366 (1966).
16. F. Klanberg and L. J. Guggenberger, *Chem. Commun.* p. 1293 (1967).
17. E. O. Fischer, K. Öfele, H. Essler, W. Fröhlich, J. P. Mortensen, and W. Semmlinger, *Z. Naturforsch.* **13b**, 458 (1958); *Ber.* **91**, 2763 (1958).

18. B. Nicholls and M. C. Whiting, *J. Chem. Soc.* p. 551 (1959).
19. G. Natta, R. Ercoli, F. Calderazzo, and S. Santambrogio, *Chim. Ind. (Milan)* **40**, 1003 (1958).
20. R. B. King and F. G. A. Stone, *J. Am. Chem. Soc.* **82**, 4557 (1960); E. O. Fischer, N. Kriebitzsch, and R. D. Fischer, *Ber.* **92**, 3214 (1959).
21. E. O. Fischer and K. Öfele, *Ber.* **91**, 2395 (1958); K. Öfele, *ibid.* **99**, 1732 (1966).
22. D. P. Tate, J. M. Augl, and W. R. Knipple, *Inorg. Chem.* **1**, 433 (1962).
23. R. B. King and A. Fronzaglia, *Inorg. Chem.* **5**, 1837 (1966).
24. H. J. Dauben, Jr. and L. R. Honnen, *J. Am. Chem. Soc.* **80**, 5570 (1958).
25. J. D. Munro and P. L. Pauson, *J. Chem. Soc.* p. 3475 (1961).
26. J. D. Munro and P. L. Pauson, *J. Chem. Soc.* p. 3479 (1961).
27. J. D. Munro and P. L. Pauson, *J. Chem. Soc.* p. 3484 (1961).
28. R. B. King and M. B. Bisnette, *Inorg. Chem.* **3**, 785 (1964).
29. R. B. King, *J. Organometal. Chem. (Amsterdam)* **8**, 129 (1967).
30. S. Winstein, H. D. Kaesz, C. G. Kreiter, and E. C. Friedrich, *J. Am. Chem. Soc.* **87**, 3267 (1965).
31. H. D. Kaesz, S. Winstein, and C. G. Kreiter, *J. Am. Chem. Soc.* **88**, 1319 (1966).
32. D. P. Tate, J. M. Augl, W. M. Ritchey, B. L. Ross, and J. G. Grasselli, *J. Am. Chem. Soc.* **86**, 3261 (1964).
33. R. B. King, *Inorg. Chem.* **7**, 1044 (1968).
34. F. A. Cotton and C. S. Kraihanzel, *J. Am. Chem. Soc.* **84**, 4432 (1962).
35a. T. S. Piper and G. Wilkinson, *J. Inorg. & Nucl. Chem.* **3**, 104 (1956).
35b. E. O. Fischer, W. Hafner, and H. O. Stahl, *Z. Anorg. Allgem. Chem.* **282**, 47 (1955).
36. G. Wilkinson, *J. Am. Chem. Soc.* **76**, 209 (1954).
37. F. C. Wilson and D. P. Shoemaker, *J. Chem. Phys.* **27**, 809 (1957).
38. R. G. Hayter, *Inorg. Chem.*, **2**, 1031 (1963).
39. R. B. King and M. B. Bisnette, *J. Organometal. Chem. (Amsterdam)* **8**, 287 (1967).
40. R. E. Dessy, R. L. Pohl, and R. B. King, *J. Am. Chem. Soc.* **88**, 5121 (1966).
41. M. Cousins and M. L. H. Green, *J. Chem. Soc.* p. 889 (1963); M. L. H. Green and A. N. Stear, *J. Organometal. Chem. (Amsterdam)* **1**, 230 (1964).
42. R. B. King and A. Fronzaglia, *J. Am. Chem. Soc.* **88**, 709 (1966).
43. R. B. King and M. B. Bisnette, *Inorg. Chem.* **4**, 486 (1965).
44. R. B. King and M. B. Bisnette, *Inorg. Chem.* **5**, 300 (1966).
45. E. W. Abel, A. Singh, and G. Wilkinson, *J. Chem. Soc.* p. 1321 (1960).
46. R. J. Haines, R. S. Nyholm, and M. H. B. Stiddard, *J. Chem. Soc.*, *A* p. 1606 (1966).
47. P. M. Treichel, K. W. Barnett, and R. L. Shubkin, *J. Organometal. Chem. (Amsterdam)* **7**, 449 (1967).
48. E. O. Fischer, K. Fichtel, and K. Öfele, *Ber.* **95**, 3172 (1962).
49. E. O. Fischer and F. J. Kohl, *Z. Naturforsch.* **18b**, 504 (1963).
50. R. B. King, *J. Am. Chem. Soc.* **85**, 1587 (1963).
51. P. M. Treichel, J. H. Morris, and F. G. A. Stone, *J. Chem. Soc.* p. 720 (1963).
52. R. B. King and M. B. Bisnette, *Inorg. Chem.* **6**, 469 (1967).
53. E. O. Fischer, O. Beckert, W. Hafner, and H. O. Stahl, *Z. Naturforsch.* **10b**, 598 (1955).
54. R. B. King, *Inorg. Chem.* **6**, 30 (1967).
55. R. B. King, *Inorg. Chem.* **7**, 90 (1968).
55a. F. A. Cotton and P. Legzdins, *J. Am. Chem. Soc.* **90**, 6232 (1968).
56. T. S. Piper and G. Wilkinson, *J. Inorg. & Nucl. Chem.* **2**, 38 (1956).
57. E. O. Fischer and P. Kuzel, *Z. Anorg. Allgem. Chem.* **317**, 226 (1962).
58. F. A. Cotton and B. F. G. Johnson, *Inorg. Chem.* **3**, 1609 (1964).
59. R. B. King and M. B. Bisnette, *Inorg. Chem.* **3**, 791 (1964).

60. E. O. Fischer and H. Strametz, *J. Organometal. Chem.* (*Amsterdam*) **10**, 323 (1967).
61. G. Wilkinson, F. A. Cotton, and J. M. Birmingham, *J. Inorg. & Nucl. Chem.* **2**, 95 (1956).
62. E. O. Fischer, K. Ulm, and H. P. Fritz, *Ber.* **93**, 2167 (1960).
63. E. O. Fischer and K. Ulm, *Ber.* **95**, 692 (1962).
64. E. O. Fischer, K. Ulm, and P. Kuzel, *Z. Anorg. Allgem. Chem.* **319**, 253 (1963).
65. M. L. H. Green, J. A. McCleverty, L. Pratt, and G. Wilkinson, *J. Chem. Soc.* p. 4854 (1961).
66. M. J. Johnson and D. F. Shriver, *J. Am. Chem. Soc.* **88**, 301 (1966).
66a. B. Deubzer and H. D. Kaesz, *J. Am. Chem. Soc.* **90**, 3276 (1968).
67. R. L. Cooper and M. L. H. Green, *J. Chem. Soc., A* p. 1155 (1967).
68. E. O. Fischer and W. Hafner, *Z. Naturforsch.* **10b**, 665 (1955).
69. E. O. Fischer, F. Scherer, and H. O. Stahl, *Ber.* **93**, 2065 (1960).
70. E. O. Fischer and H. Brunner, *Ber.* **95**, 1999 (1962); *Angew. Chem.* **76**, 577 (1964).
71. T. Kruck, *Z. Naturforsch.* **19b**, 165 (1964); H. Behrens, K. Meyer, and A. Müller, *ibid.* **20b**, 74 (1965).
72. E. O. Fischer and S. Breitschaft, *Ber.* **99**, 2213 (1966).
73. E. O. Fischer and F. J. Kohl, *Ber.* **98**, 2134 (1965).
74. R. B. King and M. B. Bisnette, *Inorg. Chem.* **3**, 785 (1964).
75. E. O. Fischer and J. Müller, *Z. Naturforsch.* **18b**, 1137 (1963).
76. E. O. Fischer and S. Breitschaft, *Ber.* **99**, 2905 (1966).
77. E. O. Fischer and S. Breitschaft, *Ber.* **99**, 2213 (1966).
78. E. Kurras and P. Klimsch, *Monatsber. Deut. Akad. Wiss. Berlin* **6**, 735 (1934).
79. W. Herwig and H. Zeiss, *J. Am. Chem. Soc.* **81**, 4891 (1959).
80. F. Hein and R. Weiss, *Z. Anorg. Allgem. Chem.* **295**, 145 (1958).
81. F. A. Cotton, *Chem. Rev.* **55**, 551 (1955).
82. H. H. Zeiss and M. Tsutsui, *J. Am. Chem. Soc.* **79**, 3062 (1957).
83. F. A. L. Anet and E. LeBlanc, *J. Am. Chem. Soc.* **79**, 2649 (1957).
84. F. A. L. Anet, *Can. J. Chem.* **37**, 58 (1959).

SUPPLEMENTARY READING

1. G. R. Dobson, I. W. Stolz, and R. K. Sheline, Substitution products of the group VIB metal carbonyls. *Advan. Inorg. Chem. Radiochem.* **8**, 1 (1966).
2. H. H. Zeiss and R. P. A. Sneeden, Sigma-bonded organochromium compounds. Hydrogen transfer reactions. *Angew. Chem. Intern. Ed. Engl.* **6**, 435 (1967).

QUESTIONS

1. Suggest reasons for the inability to prepare metal carbonyl fluorides of chromium, molybdenum, and tungsten.

2. Suggest a reason for the apparent instability of the mononuclear halides $M(CO)_5X_2$ (M = Mo or W), but the appreciable stability of their tertiary phosphine complexes such as $[(C_6H_5)_3P]_2Mo(CO)_3Cl_2$.

3. What is the function of boron trifluoride in the preparation of arene–chromium tricarbonyls from $(pyridine)_3Cr(CO)_3$ and the arene under mild conditions?

4. Suggest a reason for the observation that $(CH_3CN)_3W(CO)_3$ reacts with hexyne-2 to form a trisalkyne complex $(C_2H_5C_2C_2H_5)_3WCO$ with retention of a carbonyl group, but reacts with hexafluorobutyne-2 to form a trisalkyne complex $(CF_3C_2CF_3)_3WNCCH_3$ with retention of an acetonitrile ligand.

5. The anion $[C_5H_5Mo(CO)_3]^-$ can be obtained either from sodium cyclopentadienide and hexacarbonylmolybdenum in boiling tetrahydrofuran or from $[C_5H_5Mo(CO)_3]_2$ and

dilute sodium amalgam in tetrahydrofuran solution at room temperature. What are the major relative advantages and disadvantages of each of these preparative methods?

6. Why is it desirable to avoid large excesses of iodine in the preparation of $C_5H_5Mo(CO)_3I$ from $[C_5H_5Mo(CO)_3]_2$ and I_2?

7. Suggest a method of preparing a complex with three cyclopentadienyl rings with one ring bonded to the metal atom through all five carbon atoms, one ring bonded to the metal through three carbon atoms, and the third ring bonded to the metal through just one carbon atom.

8. Discuss the magnetic properties of $(C_5H_5)_2Cr$, $(C_6H_6)_2Cr$, $[(C_6H_6)_2Cr]^+$, $C_5H_5CrC_6H_6$, $C_5H_5CrC_7H_7$, $[C_5H_5CrC_7H_7]^+$, and $C_5H_5CrC_7H_8$. Compare them with the magnetic properties of the known analogous vanadium compounds.

9. Suggest a reason for the difference in behavior observed when the molybdenum and tungsten cations $[C_5H_5M(CO)C_6H_6]^+$ are reduced with lithium aluminum hydride.

10. Suggest a synthesis of bis(methyl benzoate)chromium.

11. Discuss the relative advantages and disadvantages of the alternate preparations of $C_5H_5CrC_7H_7$ discussed in the text.

12. Cite examples of chromium(III) organometallic derivatives with magnetic properties similar to those of conventional "inorganic" chromium(III) complexes such as the amines.

13. Compare and contrast the chemistry of hexacarbonylchromium and hexacarbonylvanadium. Attempt to account for any differences.

CHAPTER IV

Organometallic Derivatives of Manganese, Technetium, and Rhenium

Introduction

Neutral manganese, technetium, and rhenium atoms each require eleven electrons from the surrounding ligands to attain the eighteen-electron configuration of the next rare gas. Thus an $M(CO)_5$ (M = Mn, Tc, or Re) fragment would have one electron less than the rare-gas electronic configuration, and an $M(CO)_6$ (M = Mn, Tc, or Re) fragment one electron more than the rare-gas electronic configuration. However, the anions $M(CO)_5^-$ (M = Mn, Tc, or Re) and the cations $M(CO)_6^+$ (M = Mn, Tc, or Re) have the favored eighteen-electron rare-gas configuration. Both types of ions are stable species. An $M(CO)_5$ fragment can also attain the favored eighteen-electron rare-gas configuration by bonding to a one-electron donor such as a halogen or alkyl group. Thus, halides of the type $M(CO)_5X$ (M = Mn, Tc, or Re; X = Cl, Br, or I) and alkyls of the type $M(CO)_5R$ (M = Mn or Re; R = alkyl, aryl, perfluoroalkyl, acyl, etc.) are stable compounds. Two $M(CO)_5$ fragments can each attain the favored eighteen-electron rare-gas configuration by bonding together, forming an M–M covalent bond to give $M_2(CO)_{10}$ derivatives (M = Mn, Tc, or Re). These are the only stable neutral derivatives of these three metals containing only carbonyl ligands. Neutral cyclopentadienylmetal (C_5H_5M) fragments of manganese, technetium, and rhenium each require six more electrons from other ligands to attain the eighteen-electron configuration of the next rare gas. These additional six electrons can come from three carbonyl groups or one benzene ring. In the former case the very stable cyclopentadienylmetal tricarbonyls $C_5H_5M(CO)_3$ (M = Mn, Tc, or Re) are obtained. In the latter case, the cyclopentadienyl–benzene metal complexes $C_5H_5MC_6H_6$ (M = Mn or Re) are obtained; however, the preparations of these mixed sandwich compounds are extremely difficult.

A further factor in organomanganese chemistry is the unusual stability of the d^5 Mn^{2+} ion with half-filled $3d$ orbitals. Thus, the compound $(C_5H_5)_2Mn$

obtained from manganese(II) halides and sodium cyclopentadienide is the ionic manganese(II) cyclopentadienide rather than the nonionic bis-π-cyclo-pentadienylmanganese. Furthermore, the conversion of manganese(II) halides to organomanganese compounds appears to be more difficult than that of halides of neighboring transition metals to similar organometallic derivatives, apparently because of the unusual stability of Mn^{2+}. For example, the preparation of decacarbonyldimanganese $Mn_2(CO)_{10}$ from manganese(II) halides is much more difficult than the preparation of carbonyls of neighboring transition metals from their halides.

There is a great difference in the abundance of the three metals manganese, technetium, and rhenium. Manganese is among the most abundant of the transition metals and is readily available in ton quantities. Rhenium is one of the rarest elements in nature but is commercially available at a cost of about a dollar per gram. Technetium is not found in nature at all in useful quantities. However, the isotope ^{99}Tc is isolated from uranium fission products and sold for around a hundred dollars per gram. This isotope ^{99}Tc is radioactive, but its weak β-emission is stopped by ordinary glass apparatus; nevertheless, the usual precautions for radioactive materials should be taken when handling technetium. The high cost of technetium has limited the study of its organo-metallic chemistry.

The Dimetal Decacarbonyls

All three metals form fairly stable crystalline dimetal decacarbonyls $M_2(CO)_{10}$. The white rhenium compound $Re_2(CO)_{10}$ is the most stable of the three compounds, remaining unaffected when exposed to air for long periods of time. The white $Tc_2(CO)_{10}$ and yellow $Mn_2(CO)_{10}$ are slowly oxidized by air in the solid state over a period of weeks and in solution over a period of hours. However, $Mn_2(CO)_{10}$ is sufficiently stable and volatile to be purified by steam distillation.

Crystal structure studies by x-ray diffraction have shown the $M_2(CO)_{10}$ derivatives to have structure I (M = Mn, Tc, or Re) with a covalent metal–metal bond as the sole link between the two halves of the molecule [1]. The metal–metal distances in these $M_2(CO)_{10}$ compounds are in the vicinity of 3 Å.

I

Decacarbonyldimanganese $Mn_2(CO)_{10}$ is among the most difficult of the simple stable metal carbonyls to prepare efficiently. Most methods which work well for the preparation of other metal carbonyls fail when applied to deca-carbonyldimanganese. Reduction of manganese(II) iodide with magnesium metal in diethyl ether in the presence of carbon monoxide under pressure sometimes gives traces ($<1\%$) of decacarbonyldimanganese; sufficient $Mn_2(CO)_{10}$ for proper characterization was obtained by this method [2]. Reaction of manganese(II) chloride with sodium benzophenone ketyl followed by treatment of the resulting manganese ketyl solution with carbon monoxide under pressure gave sufficient $Mn_2(CO)_{10}$) ($\sim30\%$ yield) for a detailed study of its chemistry [3]. Somewhat better yields (50–60%) of $Mn_2(CO)_{10}$ can be obtained by treatment of a mixture of manganese(II) acetate, diisopropyl ether, and excess triethylaluminum or triisobutylaluminum with carbon monoxide under pressure [4]. Although trialkylaluminum compounds are readily available, their handling in the large quantities required for the preparation of useful quantities of $Mn_2(CO)_{10}$ is very hazardous.

Despite the difficulties associated with the preparation of $Mn_2(CO)_{10}$, this carbonyl derivative is, at present, readily available. π-Methylcyclopenta-dienyltricarbonylmanganese $CH_3C_5H_4Mn(CO)_3$ is manufactured in tonnage quantities for use as a combustion improver of fuel oils and, therefore, is one of the most inexpensive metal carbonyl derivatives commercially available at the present time. It is possible to convert $CH_3C_5H_4Mn(CO)_3$ to $Mn_2(CO)_{10}$ by reaction with sodium metal and carbon monoxide in diglyme solution at elevated temperatures (even at atmospheric pressure) followed by cautious acid hydrolysis of the resulting reaction mixture [5].

The dimetal decacarbonyls of technetium and rhenium are most con-veniently prepared by reactions of their oxides (e.g., Re_2O_7 or TcO_2) or their oxyacid salts (e.g., $KReO_4$) with carbon monoxide at elevated temperatures and pressures, e.g.:

$$Re_2O_7 + 17\,CO \longrightarrow Re_2(CO)_{10} + 7\,CO_2 \qquad [6] \qquad\qquad (a)$$

$$2\,TcO_2 + 14\,CO \longrightarrow Tc_2(CO)_{10} + 4\,CO_2 \qquad [7] \qquad\qquad (b)$$

Decacarbonyldirhenium, $Re_2(CO)_{10}$, has also been prepared by reaction of rhenium pentachloride with sodium metal and carbon monoxide in diglyme solution at elevated pressures [8].

Other Metal Carbonyl Derivatives

The dimetal decacarbonyls of manganese, technetium, and rhenium can be used for the preparation of a variety of other carbonyl derivatives of these metals. Most of these carbonyl derivatives are of the type $M(CO)_5Y$, where Y is a one-electron donor, and therefore, have the favored eighteen-electron rare-gas configuration.

Much of the preparative chemistry of manganese and rhenium carbonyls utilizes the pentacarbonylmetallate anions $M(CO)_5^-$ as reactive intermediates. These anions are most conveniently prepared by reduction of $M_2(CO)_{10}$ (M = Mn or Re) with a ~1% sodium amalgam in tetrahydrofuran solution according to the following equation [9]:

$$M_2(CO)_{10} + 2\,Na \longrightarrow 2\,NaM(CO)_5$$

The resulting solutions of $NaM(CO)_5$ salts react with a variety of organic and organometallic halides to form nonionic $RM(CO)_5$ compounds according to the following equation:

$$NaM(CO)_5 + RX \longrightarrow RM(CO)_5 + NaX$$

Halides such as methyl iodide, benzyl chloride, and a large number of acyl chlorides form white to yellow air-stable derivatives $RM(CO)_5$ with metal–carbon σ-bonds. Compounds with manganese–tin, manganese–gold, and manganese–iron bonds may be obtained by reaction of $NaMn(CO)_5$ with the halides R_3SnCl (R = methyl or phenyl), $(C_6H_5)_3PAuCl$, and $C_5H_5Fe(CO)_2I$, respectively. Treatment of the sodium salts $NaM(CO)_5$ with a strong non-volatile, nonoxidizing acid such as phosphoric acid gives the volatile liquid "hydrides" $HM(CO)_5$ (**II**: M = Mn or Re; the less stable technetium derivative is obtained only in trace quantities). The $HM(CO)_5$ derivatives are weak acids, stable thermally to about 100°, but oxidized rapidly by air to $M_2(CO)_{10}$. The volatility of the $HM(CO)_5$ compounds is sufficient that they can be manipulated by high-vacuum techniques.

OC, OC, M, CO, CO, H, with central carbonyl C≡O

II

Some $RM(CO)_5$ derivatives undergo a reversible carbonylation reaction which may be related to processes occurring in carbonylation reactions of organic compounds catalyzed by metal carbonyls [10]. Treatment of $NaMn(CO)_5$ with methyl iodide gives the methyl derivative $CH_3Mn(CO)_5$, mp 95°. This methyl derivative reacts with carbon monoxide at room temperature and atmospheric pressure to give the acetyl derivative $CH_3COMn(CO)_5$, mp 54–55°, which is identical to the product obtained from $NaMn(CO)_5$ and acetyl chloride. Furthermore, this carbonylation reaction is reversible. Thus, heating the acetyl derivative $CH_3COMn(CO)_5$ to 80° causes the loss of one mole of carbon monoxide, yielding the methyl derivative $CH_3Mn(CO)_5$. Studies with labeled carbon monoxide have shown that the carbon monoxide lost in this decarbonylation reaction does not come from the acetyl carbonyl group (that in CH_3CO) but, instead, from the terminal carbonyl group [that in $Mn(CO)_5$]; thus, during this decarbonylation reaction the acetyl carbonyl group becomes a terminal carbonyl group.

The number of different acyl halides which react with $NaMn(CO)_5$ to form $R'COMn(CO)_5$ is much larger than the number of different alkyl halides which react with $NaMn(CO)_5$ to form $RMn(CO)_5$ compounds. For example, the reactivity of iodobenzene is so low that it does not react with $NaMn(CO)_5$ to give the phenyl derivative $C_6H_5Mn(CO)_5$. However, the much more reactive benzoyl chloride reacts with $NaMn(CO)_5$ to give a good yield of the benzoyl derivative $C_6H_5COMn(CO)_5$. Upon heating this benzoyl derivative, smooth elimination of carbon monoxide occurs, giving a good yield of the phenyl derivative $C_6H_5Mn(CO)_5$. A similar technique is useful for preparing the perfluoroalkyl derivatives $R_fMn(CO)_5$ [$R_f = CF_3$, C_2F_5, C_3F_7, $(CF_3)_2CF$, $\frac{1}{2}—CF_2CF_2CF_2—$, C_6F_5, etc.]. Reaction of $NaMn(CO)_5$ with the perfluoroacyl chlorides, R_fCOCl, gives the white perfluoroacyl derivatives $R_fCOMn(CO)_5$. Heating these derivatives to $80°–100°$ results in the elimination of one mole of carbon monoxide to give the corresponding perfluoroalkyl derivatives $R_fMn(CO)_5$ [11]. A few rhenium analogues $R_fRe(CO)_5$ have also been prepared by similar methods, but using decarbonylation temperatures $30°–40°$ higher than those used for the manganese analogues, in accord with the greater stability of rhenium–carbon bonds.

Other preparative methods are sometimes useful for the preparation of fluoroalkylmetal pentacarbonyls. An important reaction is the addition of $HMn(CO)_5$ to the carbon–carbon double bond of a highly fluorinated olefin (hydrometallation). This technique is illustrated by the following preparation of the tetrafluoroethyl derivative $HCF_2CF_2M(CO)_5$ [12]:

$$HM(CO)_5 \; + \; CF_2{=\!=}CF_2 \quad \xrightarrow{25°} \quad HCF_2CF_2M(CO)_5$$

Another class of carbonyl derivatives of manganese, technetium, and rhenium includes those with halogen ligands, such as the metal carbonyl halides. Both the monometallic derivatives $M(CO)_5X$ and the bimetallic derivatives $[M(CO)_4X]_2$ (III) are known [13]; the metal atoms in both of these compound types possess the favored eighteen-electron rare-gas configuration. The bimetallic compounds $[M(CO)_4X]_2$ have two halogen bridges holding together the two halves of the molecule. Heating the monometallic derivatives $M(CO)_5X$ to about $100°$ removes one mole of carbon monoxide, giving the bimetallic derivatives $[M(CO)_4X]_2$ (III). Conversely, reaction of the bimetallic derivatives $[M(CO)_4X]_2$ (III) with carbon monoxide cleaves the halogen bridges giving the monometallic $M(CO)_5X$.

$$
\begin{array}{c}
\quad\quad \overset{\displaystyle O}{\overset{\displaystyle \|}{\underset{}{C}}} \quad\quad\quad \overset{\displaystyle O}{\overset{\displaystyle \|}{\underset{}{C}}} \\
OC\diagdown \quad | \quad \diagup X \diagdown \quad | \quad CO \\
\quad\quad M \quad\quad\quad M \\
OC\diagup \quad | \quad \diagdown X \diagup \quad | \quad CO \\
\quad\quad \overset{}{\underset{\displaystyle O}{\underset{\displaystyle \|}{C}}} \quad\quad\quad \overset{}{\underset{\displaystyle O}{\underset{\displaystyle \|}{C}}}
\end{array}
$$

III

Most of the metal carbonyl halides $M(CO)_5X$ can be obtained by cleaving the metal–metal bond in $M_2(CO)_{10}$ with halogens, i.e.,

$$M_2(CO)_{10} + X_2 \longrightarrow 2\,M(CO)_5X$$

where M = Mn, Tc, or Re; X = Cl or Br. This reaction occurs readily with chlorine or bromine at room temperature in an inert solvent such as carbon tetrachloride or dichloromethane. However, iodine requires heating to react with $M_2(CO)_{10}$. A better route to the iodides $M(CO)_5I$ utilizes the reaction between the more reactive sodium salt $NaM(CO)_5$ and iodine in tetrahydrofuran, according to the following equation:

$$NaM(CO)_5 + I_2 \longrightarrow M(CO)_5I + NaI$$

The technetium and rhenium carbonyl halides can be prepared by cleaving the corresponding carbonyl $M_2(CO)_{10}$ as discussed above. The need for the carbonyls $M_2(CO)_{10}$ can be bypassed by preparing the $M(CO)_5X$ halides (M = Tc or Re) by carbonylation of a mixture of the appropriate dipotassium hexahalometallate(IV) K_2MX_6 and excess copper powder [14].

Two manganese carbonyl nitrosyls are known: red $Mn(CO)_4NO$, mp 0°, and green $MnCO(NO)_3$, mp 27°. Both compounds are very volatile. The red $Mn(CO)_4NO$ was first obtained by reaction of $HMn(CO)_5$ with N-methyl-N-nitroso-p-toluenesulfonamide (Diazald) in diethyl ether solution at room temperature [15]. A better preparative method utilizes the reaction of the triphenylphosphine complex $[(C_6H_5)_3PMn(CO)_4]_2$ with nitric oxide in Tetralin (1,2,3,4-tetrahydronaphthalene) at 95° according to the following equation [16]:

$$[(C_6H_5)_3PMn(CO)_4]_2 + 2\,NO \longrightarrow$$
$$Mn(CO)_4NO + (C_6H_5)_3PMn(CO)_3NO + CO + (C_6H_5)_3P$$

The green $MnCO(NO)_3$ was first obtained in very low yield by reaction of $Mn(CO)_5I$ with nitric oxide at 90°–100° [17]. A much better method for the preparation of $MnCO(NO)_3$ utilizes the reaction between $Mn(CO)_4NO$ and nitric oxide in xylene solution at 90° according to the following equation [16]:

$$Mn(CO)_4NO + 2\,NO \longrightarrow MnCO(NO)_3 + 3\,CO$$

The carbonyl groups in $Mn(CO)_4NO$ and $MnCO(NO)_3$ can be replaced by tertiary phosphines to give products which are more stable than the parent compounds.

The mononuclear metal carbonyl hydrides $HM(CO)_5$ ($M = Mn$ or Re) were discussed above. Some trinuclear carbonyl hydrides of manganese, technetium, and rhenium have also been prepared by techniques exemplified by the following reactions [18]:

$$Re_2(CO)_{10} \xrightarrow[\text{THF}/65°]{\text{NaBH}_4} \xrightarrow{\text{H}_3\text{PO}_4} [HRe(CO)_4]_3$$
$$\text{(white air-stable solid)}$$

$$Re_2(CO)_{10} \xrightarrow[\text{THF}/65°]{\text{NaBH}_4} \xrightarrow{\text{NaRe(CO)}_5} \xrightarrow{\text{H}_3\text{PO}_4} HRe_3(CO)_{14}$$
$$[\text{NMR}: \tau \, (\text{Re}-\underline{\text{H}}) = 26.25]$$

$$Re_2(CO)_{10} \xrightarrow[\text{THF}/65°]{\text{NaBH}_4} \xrightarrow{\text{NaMn(CO)}_5} \xrightarrow{\text{H}_3\text{PO}_4} HRe_2Mn(CO)_4$$

The compounds $[HM(CO)_4]_3$ ($M = Mn$, Tc, or Re) have a closed triangular array of the three metal atoms, as in structure **IV**. However, the compounds $HM_2M'(CO)_{14}$ (M and $M' = Mn$ or Re) have an open bent array of the three metal atoms, as in structure **V**. Reaction of $Mn_2(CO)_{10}$ with sodium borohydride in tetrahydrofuran at 65° gives the red, volatile boron–manganese derivative $HB_2H_6Mn_3(CO)_{10}$ (**VI**) [19].

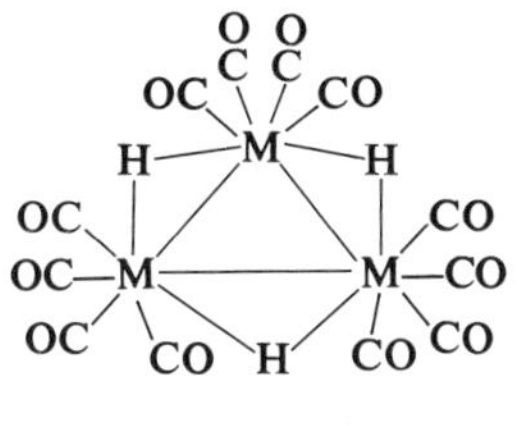

IV

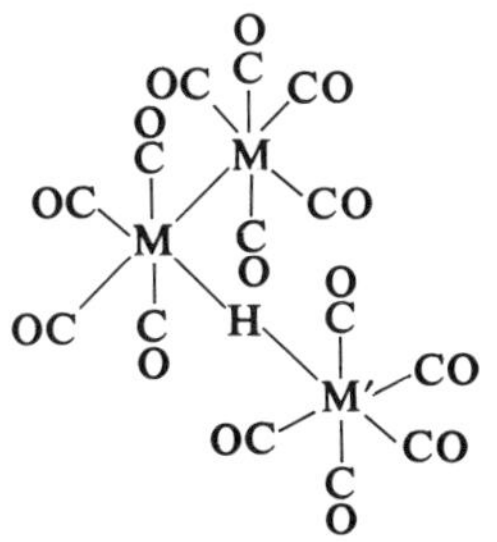

V

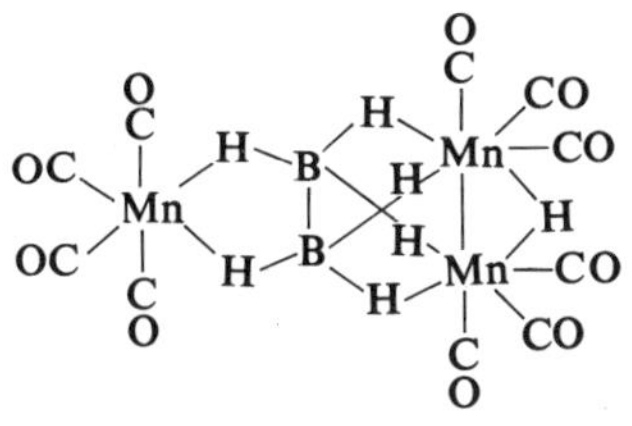

VI

π-Cyclopentadienyl Derivatives

Reaction of anhydrous manganese(II) bromide with sodium cyclopenta-dienide in tetrahydrofuran solution gives dark brown, volatile, air-sensitive, antiferromagnetic $(C_5H_5)_2Mn$, mp 158°. This compound is formulated as the ionic manganese(II) cyclopentadienide rather than the nonionic bis-π-cyclo-pentadienylmanganese(II) for the following reasons [20]. (1) Water reacts rapidly with $(C_5H_5)_2Mn$ to give cyclopentadiene and manganese(II) oxide; by contrast, authentic π-cyclopentadienyl derivatives such as $(C_5H_5)_2Fe$ do not react with water. (2) A tetrahydrofuran solution of $(C_5H_5)_2Mn$ reacts rapidly with ferrous chloride to give a good yield of ferrocene. Authentic π-cyclopentadienyl derivatives do not form ferrocene upon treatment with ferrous chloride. (3) Above the transition temperature of 158°, the dark brown $(C_5H_5)_2Mn$ becomes colorless and loses its antiferromagnetism. At this point the magnetic moment of $(C_5H_5)_2Mn$ is 5.9 BM, corresponding to the five unpaired electrons characteristic of Mn^{2+}. The ionic nature of $(C_5H_5)_2Mn$ is a further indication of the stability of the Mn^{2+} ion with its half-filled shell.

Rhenium and technetium do not form $(C_5H_5)_2M$ derivatives. Reaction of rhenium pentachloride with excess sodium cyclopentadienide gives yellow, volatile crystals of $(C_5H_5)_2ReH$ (**VII**: M = Re) [21]. As might be expected, the yield of this rhenium hydride derivative is greatly improved upon adding sodium borohydride to the reaction mixture. Much more recently, a similar technetium compound $(C_5H_5)_2TcH$ (**VII**: M = Tc) was prepared by reaction between technetium tetrachloride, sodium cyclopentadienide, and sodium borohydride [22]. This technetium compound appears to be identical to a species previously [23] reported as $[(C_5H_5)_2Tc]_2$. The hydrides $(C_5H_5)_2MH$ (M = Tc or Re) have the favored eighteen-electron rare-gas configuration and exhibit the expected diamagnetism.

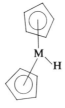

VII

A variety of reactions of the rhenium compound $(C_5H_5)_2ReH$ (**VII**: M = Re) have been investigated. The hydrogen in $(C_5H_5)_2ReH$ has no detectable acidic properties. However, the water-insoluble $(C_5H_5)_2ReH$ dissolves in aqueous acids to form colorless solutions containing the cation $[(C_5H_5)_2ReH_2]^+$; solid salts of this cation have been isolated. The hydride $(C_5H_5)_2ReH$ thus possesses basic properties; the base strength of $(C_5H_5)_2ReH$ is similar to that of ammonia. Reaction of the cation $[(C_5H_5)_2ReH_2]^+$ with aqueous base regenerates the parent hydride $(C_5H_5)_2ReH$.

In an attempt to prepare cyclopentadienylrhenium carbonyls, the reaction of $(C_5H_5)_2ReH$ with carbon monoxide was carried out at $100°/250$ atm. Under these conditions a pale yellow product $C_5H_5Re(CO)_2C_5H_6$ was obtained [24]. Proton NMR studies on this product indicated it to have structure **VIII** with a cyclopentadiene ligand bonded to the rhenium atom by means of one double bond. The uncomplexed double bond in $C_5H_5Re(CO)_2C_5H_6$ (**VIII**) can be hydrogenated in the presence of a platinum catalyst to give the cyclopentene complex $C_5H_5Re(CO)_2C_5H_8$ (**IX**). Reaction of $(C_5H_5)_2ReH$ with the acetylenic esters $RC\equiv CCO_2CH_3$ ($R = H$ or CO_2CH_3) in tetrahydrofuran solution at room temperature results in *cis*-addition to the triple bond, giving dark red *cis*-$(C_5H_5)_2Re$—$CR\equiv CHCO_2CH_3$ (**X**) [25]. Metallation of $(C_5H_5)_2ReH$ with butyllithium followed by treatment with methyl iodide gives yellow $(C_5H_5)(CH_3C_5H_5)Re(CH_3)_2$, mp 73–74°. This compound has structure **XI** containing two methyl groups directly bonded to the rhenium atom and a methylcyclopentadiene ligand behaving as a π-bonded diene [26]. Apparently, butyllithium reacts with $(C_5H_5)_2ReH$ both by replacing the hydrogen atom bonded to the rhenium and by metallating one of the cyclopentadienyl rings. Halogens react with $(C_5H_5)_2ReH$ to give salts of the $[(C_5H_5)_2ReX]^+$ cations ($X = Cl$, Br, or I) [27].

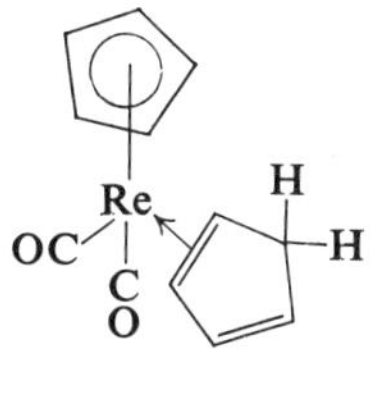

VIII

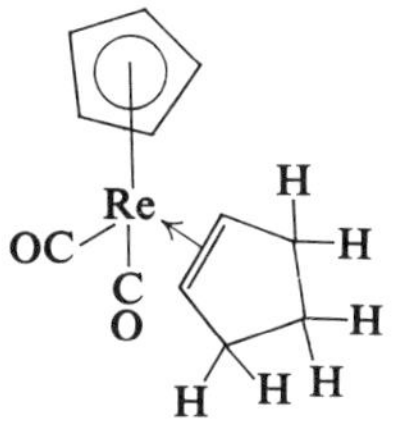

IX

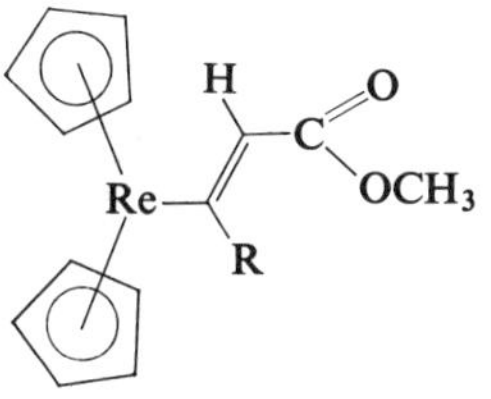

X

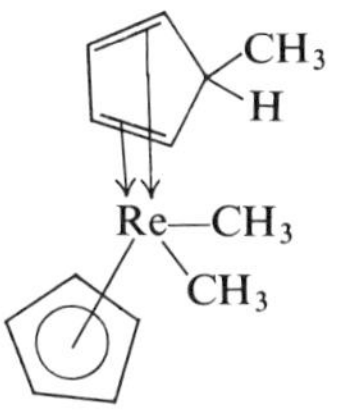

XI

Some manganese and rhenium compounds containing both π-bonded cyclopentadienyl and benzene rings have been obtained. Reaction of manganese(II) chloride with a mixture of sodium cyclopentadienide and phenyl-magnesium bromide in tetrahydrofuran solution followed by hydrolysis gives a mixture of red π-benzene-π-cyclopentadienylmanganese $C_5H_5MnC_6H_6$ (**XII**: M = Mn, mp 191°–192° (2% yield); red π-biphenyl-π-cyclopentadienyl-manganese $C_5H_5MnC_6H_5C_6H_5$ (**XIII**), mp 74–76° (15% yield); and red π-biphenyl-bis(π-cyclopentadienylmanganese) $C_5H_5MnC_6H_5C_6H_5MnC_5H_5$ (**XIV**), mp 194°–195° (3% yield) [28]. Ultraviolet irradiation of a mixture of rhenium pentachloride, cyclopentadienylmagnesium bromide, and 1,3-cyclo-hexadiene in diethyl ether solution gives a low yield of yellow π-benzene-π-cyclopentadienylrhenium $C_5H_5ReC_6H_6$ (**XII**: M = Re). This water-insoluble rhenium complex **XII** (M = Re), like the hydride $(C_5H_5)_2ReH$ (**VII**: M = Re), dissolves in aqueous acids. Protonation occurs to form the $[C_5H_5ReHC_6H_6]^+$ (**XV**) cation [29]. All of these manganese and rhenium complexes with both π-bonded cyclopentadienyl and π-bonded benzene rings have the favored eighteen-electron rare-gas configuration.

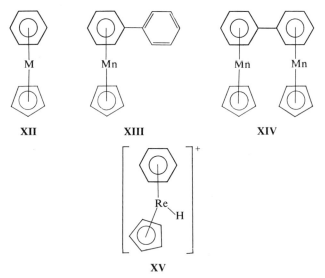

XII   XIII   XIV

XV

Manganese, technetium, and rhenium all form very stable cyclopentadienyl-metal tricarbonyls of the general formula $C_5H_5M(CO)_3$ (**XVI**: R = H;

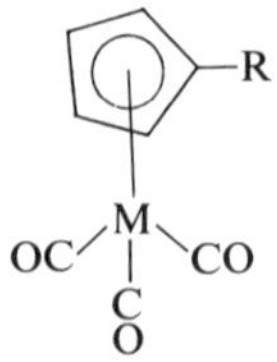

XVI

M = Mn, Tc, or Re) which possess the favored eighteen-electron rare-gas configuration. The manganese compound (known colloquially as cymantrene from *cyclopentadienylman*ganese *tri*carbonyl) is a yellow crystalline solid. The rhenium and technetium analogues are white solids consistent with the general trend toward lighter colors of organometallic compounds upon descending a column of the periodic table.

The original preparation of $C_5H_5Mn(CO)_3$ utilized the reaction of $(C_5H_5)_2Mn$ with carbon monoxide, preferably at elevated pressures and temperatures [30]. However, since the methyl derivative $CH_3C_5H_4Mn(CO)_3$ (**XVI**: R = CH_3; M = **Mn**) has become commercially important as a combustion improver in fuel oils, considerable effort has been expended, largely by workers at the Ethyl Corporation, in developing efficient methods for converting inexpensive manganese compounds to $C_5H_5Mn(CO)_3$ or its methyl derivative $CH_3C_5H_4Mn(CO)_3$. An illustration of such an improved technique utilizes the reaction between the pyridine complex $(C_5H_5N)_2MnCl_2$, magnesium, cyclopentadiene, and carbon monoxide under pressure in dimethylformamide solution in the presence of hydrogen for the preparation of $C_5H_5Mn(CO)_3$ in up to 80% yield [31]. The technetium and rhenium compounds $C_5H_5M(CO)_3$ (**XVI**: R = H, M = Tc or Re) are best prepared by treatment of the corresponding metal pentacarbonyl chloride $M(CO)_5Cl$ with sodium cyclopentadienide in boiling tetrahydrofuran according to the following equation [32]:

$$M(CO)_5Cl + NaC_5H_5 \xrightarrow{\text{THF}} C_5H_5M(CO)_3 + 2\ CO + NaCl$$

where M = Tc or Re. The required halides $M(CO)_5Cl$ (M = Tc or Re) are prepared by reactions of the readily available K_2MCl_6 salts (M = Tc or Re) with carbon monoxide at elevated temperatures and pressures in the presence of excess copper powder as a reducing agent.

The reactions of $C_5H_5Mn(CO)_3$ have been extensively investigated. Known reactions are of the following two types: (1) substitution on the π-cyclopentadienyl ring, (2) replacement of carbonyl groups.

SUBSTITUTION ON THE π-CYCLOPENTADIENYL RING

The π-cyclopentadienyl ring in $C_5H_5Mn(CO)_3$ can undergo substitution with electrophilic reagents in a manner similar to benzene and ferrocene. The extensive involvement of the three carbonyl groups of $C_5H_5Mn(CO)_3$ in retrodative π-bonding withdraws electrons from the antibonding E_2 orbitals of the π-C_5H_5 ring, decreasing the electron density on the C_5H_5 ring and, therefore, reducing its reactivity toward electrophilic reagents. Thus, the reactivity of $C_5H_5Mn(CO)_3$ toward electrophilic substitution is less than that of ferrocene. Furthermore, the manganese compound $C_5H_5Mn(CO)_3$ undergoes a lesser variety of electrophilic substitution reactions than ferrocene. Treatment of $C_5H_5Mn(CO)_3$ with acyl chlorides in the presence of aluminum chloride results in the formation of acyl derivatives according to the following equation [33]:

$$C_5H_5Mn(CO)_3 + RCOCl \xrightarrow{AlCl_3} RCOC_5H_4Mn(CO)_3 + HCl$$

where $R = CH_3$, C_6H_5, etc. These ketones are useful intermediates for the preparation of other substituted cyclopentadienylmanganese tricarbonyls. Thus, the compounds $RCOC_5H_4Mn(CO)_3$ (**XVII**) react with alkylmagnesium halides R'MgX in diethyl ether solution to give, after hydrolysis, the alcohols $RR'C(OH)C_5H_4Mn(CO)_3$. These alcohols can be easily dehydrated to olefins in cases where either the R or R' groups have a hydrogen on the α-carbon atom. Reaction of $C_5H_5Mn(CO)_3$ with sulfuric acid in acetic anhydride solution gives the sulfonic acid $(CO)_3MnC_5H_4SO_3H$ (**XVIII**) which can be isolated as its p-toluidinium salt [34]. This sulfonic acid **XVIII** reacts with mercuric chloride with replacement of the sulfonic acid group by a chloromercuri group to give $ClHgC_5H_4Mn(CO)_3$ (**XIX**).

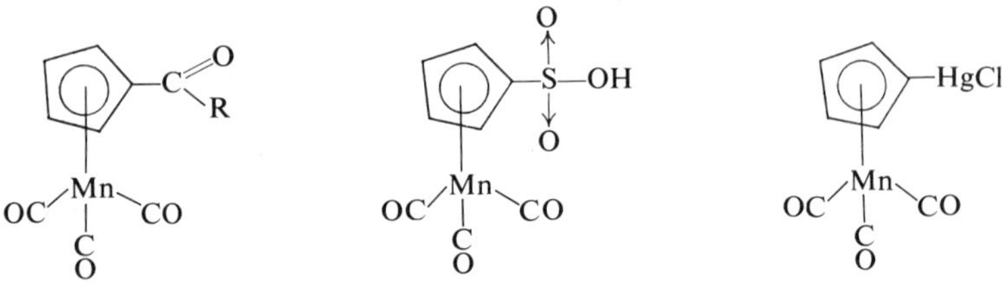

XVII   XVIII   XIX

REPLACEMENT OF CARBONYL GROUPS

Reaction of $C_5H_5Mn(CO)_3$ with various Lewis bases or other electron-donor ligands can result in the replacement of one (occasionally more) carbonyl group with another ligand, giving a $C_5H_5Mn(CO)_2L$ derivative. In most cases ultraviolet irradiation is necessary to cause such reactions to take place, mere heating being insufficient.

The following reactions illustrate substitution reactions of $C_5H_5Mn(CO)_3$ which take place by ultraviolet irradiation at room temperature in solutions of inert solvents [35]:

$$C_5H_5Mn(CO)_3 + L \xrightarrow{UV} C_5H_5Mn(CO)_2L + CO$$

where L = amines, phosphines, sulfoxides, olefins, and isonitriles;

$$C_5H_5Mn(CO)_3 + C_4H_6 \xrightarrow{UV} C_5H_5Mn(CO)C_4H_6 + 2CO$$

(butadiene)   (XX)

$$C_5H_5Mn(CO)_3 + diphos \xrightarrow{UV} C_5H_5Mn(CO)(diphos) + 2CO$$

where diphos = a chelating ditertiary phosphine such as

$$(C_6H_5)_2PCH_2CH_2P(C_6H_5)_2$$

Sometimes, bridged derivatives of the type $(diphos)[Mn(CO)_2C_5H_5]_2$ are also formed.

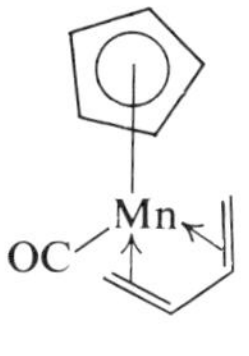

XX

A further illustration of the replacement of carbonyl groups is the preparation of the nitrosyl cation $[C_5H_5Mn(CO)_2NO]^+$ (XXI) by heating $C_5H_5Mn(CO)_3$ with sodium nitrite and hydrochloric acid in ethanol solution according to the following equation [36]:

$$C_5H_5Mn(CO)_3 + NaNO_2 + 2HCl \xrightarrow{ethanol} [C_5H_5Mn(CO)_2NO]^+ + NaCl + H_2O + Cl^-$$

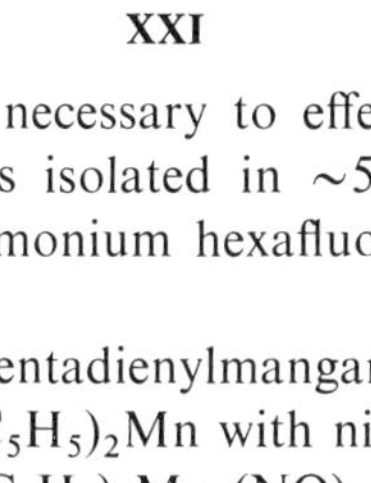

XXI

Ultraviolet irradiation is not necessary to effect this reaction. The yellow cation $[C_5H_5Mn(CO)_2NO]^+$ is isolated in ~50% yield as its hexafluorophosphate by addition of ammonium hexafluorophosphate to the reaction mixture.

A variety of unusual cyclopentadienylmanganese nitrosyl derivatives have been prepared. Reaction of $(C_5H_5)_2Mn$ with nitric oxide in tetrahydrofuran solution gives purple-black $(C_5H_5)_3Mn_2(NO)_3$ of uncertain structure [37]. Other cyclopentadienylmanganese nitrosyl derivatives are obtained by reactions of the cation $[C_5H_5Mn(CO)_2NO]^+$ with nucleophiles. Thus, ultra-violet irradiation of $[C_5H_5Mn(CO)_2NO]^+$ with aqueous sodium nitrite [38] gives black, sparingly soluble $[C_5H_5Mn(NO)_2]_n$ of uncertain molecular weight. Reaction of $[C_5H_5Mn(CO)_2NO]^+$ with sodium methoxide in methanol solution gives the red crystalline methoxycarbonyl derivative $C_5H_5Mn(CO)(NO)(CO_2CH_3)$ (**XXII**: $Y = OCH_3$) which reacts with methyl-magnesium bromide to give, after hydrolysis, the orange-brown, unstable liquid acetyl derivative $C_5H_5Mn(CO)(NO)(COCH_3)$ (**XXII**: $Y = CH_3$) [39].

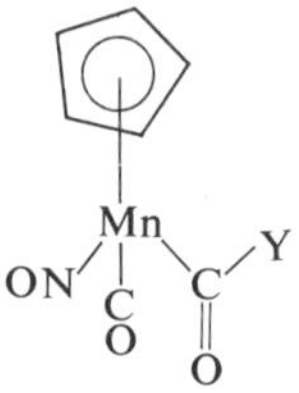

XXII

These compounds are the only known $RMn(CO)(NO)(C_5H_5)$ derivatives. Reduction of $[C_5H_5Mn(CO)_2NO]^+$ (**XXI**) with aqueous sodium borohydride gives the red-purple $[C_5H_5Mn(CO)(NO)]_2$, indicated by its infrared spectrum in the $\nu(CO)$ and $\nu(NO)$ regions to have all four of the following pos-sibilities: terminal carbonyl groups, terminal nitrosyl groups, bridging carbonyl groups, and bridging nitrosyl groups [36]. Ultraviolet irradiation of $[C_5H_5Mn(CO)(NO)]_2$ gives the black trinuclear complex $(C_5H_5)_3Mn_3(NO)_4$ shown by x-ray crystallography [40] to have structure **XXIII** with three two-way bridging nitrosyl groups and one three-way bridging nitrosyl group.

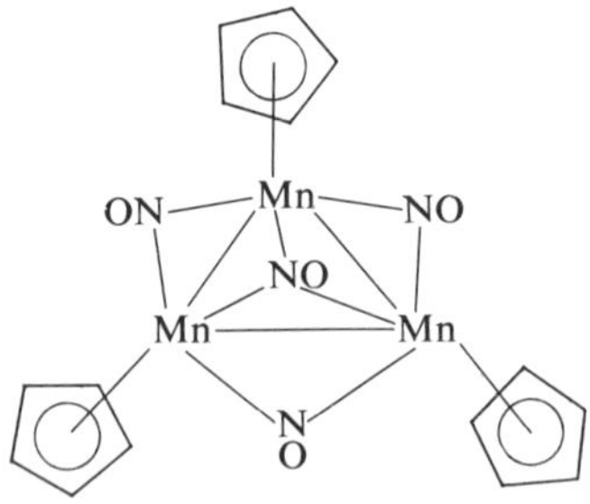

XXIII

Miscellaneous Organometallic Derivatives

π-Pyrrolylmanganese Carbonyls [41]

The pyrrolide anion $C_4H_4N^-$ is isoelectronic with the cyclopentadienide anion $C_5H_5^-$ which forms many stable metal π-complexes. Attempts to prepare π-pyrrolyl derivatives analogous to π-cyclopentadienyl derivatives rarely succeed. The tendency for a C_4H_4N ligand to π-bond to a transition metal thus appears to be less than the tendency for a C_5H_5 ligand to π-bond to a transition metal. Despite these difficulties the yellow-orange π-pyrrolyl-manganese tricarbonyl complex $C_4H_4NMn(CO)_3$ (**XXIV**), mp 41°, iso-electronic with $C_5H_5Mn(CO)_3$ (**XVI**: R = H; M = Mn), has been prepared, either by heating $Mn_2(CO)_{10}$ with pyrrole in ligroin solution at 100° or by reaction of $Mn(CO)_5Br$ with potassium pyrrolide in boiling tetrahydrofuran. Similar pyrrolylmanganese compounds have been prepared from substituted pyrroles.

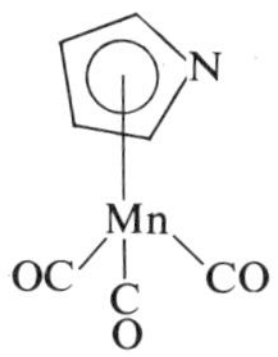

XXIV

π-BENZENE AND π-CYCLOHEXADIENYL COMPLEXES OF MANGANESE CARBONYLS

Treatment of $Mn(CO)_5Br$ in benzene solution with anhydrous aluminum chloride at the boiling point gives a mixture from which the cation $[C_6H_6Mn(CO)_3]^+$ (**XXV**) can be isolated as its perchlorate (explosion danger!) or hexafluorophosphate [42]. The cation $[C_6H_6Mn(CO)_3]^+$ is isoelectronic

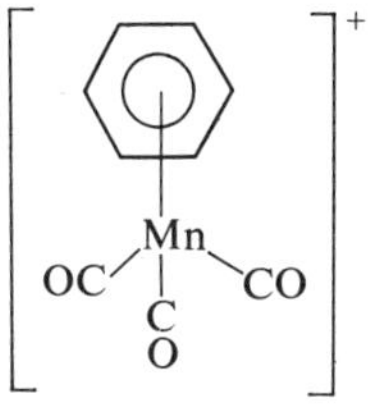

XXV 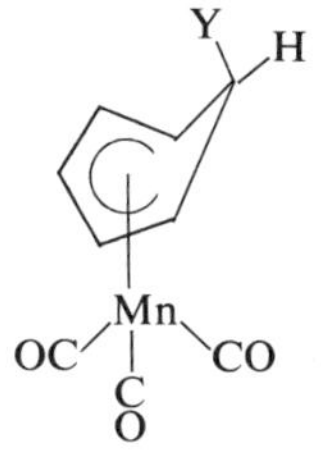   **XXVI**

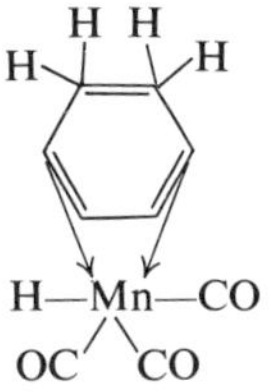

XXVII

with the neutral $C_6H_6Cr(CO)_3$ (p. 68). Dry salts of this $[C_6H_6Mn(CO)_3]^+$ cation are reported to give the following products upon reduction with lithium aluminum hydride in tetrahydrofuran solution: (a) the π-cyclohexadienyl complex $C_6H_7Mn(CO)_3$ (**XXVI**: Y = H [this compound can also be obtained in low yield by heating $Mn_2(CO)_{10}$ with 1,3-cyclohexadiene]; (b) the π-cyclo-hexadiene-manganese carbonyl hydride $C_6H_8Mn(CO)_3H$ (**XXVII**) [43]. This product obtained in low yield, is similar to the compound $C_5H_5WC_6H_8(CO)H$ obtained by the $LiAlH_4$ reduction of $[C_5H_5W(CO)C_6H_6]^+$ (p. 87). Reaction of the $[C_6H_6Mn(CO)_3]^+$ cation with other nucleophiles gives substituted π-cyclohexadienyl derivatives; thus the reaction of $[C_6H_6Mn(CO)_3]^+$ with methyllithium gives *exo*-π-methylcyclohexadienylmanganese tricarbonyl $CH_3C_6H_6Mn(CO)_3$ (**XXVI**: Y = CH_3) [44].

BIS(HEXAMETHYLBENZENE) AND RELATED COMPLEXES OF RHENIUM [42]

Arene and cyclohexadienyl derivatives of rhenium similar to the manganese compounds discussed above have been prepared, but with six methyl substituents on the six-membered rings and with a π-hexamethylbenzene ring replacing the three carbonyl groups. Reaction of rhenium trichloride with hexamethylbenzene, aluminum chloride, and aluminum metal at elevated temperatures followed by hydrolysis and addition of ammonium hexafluorophosphate gives the colorless salt $\{[(CH_3)_6C_6]_2Re\}[PF_6]$. Reduction of this salt can be carried out in two different ways. Thus, treatment of this salt with lithium aluminum hydride in tetrahydrofuran solution gives the yellow complex $(CH_3)_6C_6ReC_6H(CH_3)_6$ with structure **XXVIII**, containing a π-hexamethylcyclohexadienyl ring [45]. However, reduction of the bis(hexamethylbenzene)rhenium cation with molten sodium in the absence of a solvent or other source of protons gives bis(hexamethylbenzene)rhenium dimer

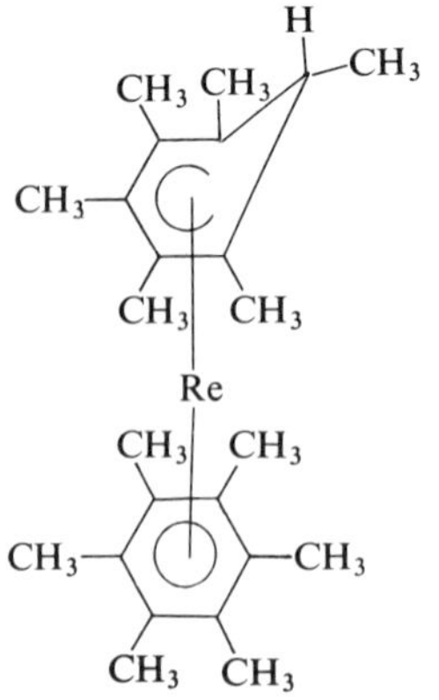

XXVIII 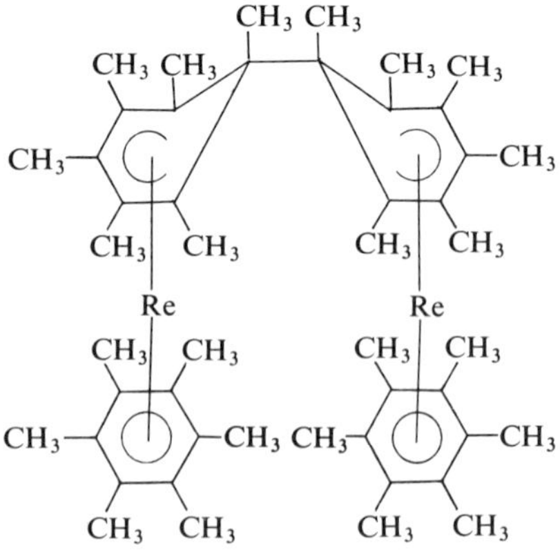   XXIX

$\{[(CH_3)_6C_6]_2Re\}_2$ of apparent structure **XXIX** with a carbon–carbon bond holding the two halves of the molecule together. This carbon–carbon bond is relatively easily broken. Heating **XXIX** in a vacuum and condensing the vapors on a $-196°$ probe gives appreciable amounts of monomeric bis(hexamethylbenzene)rhenium $[(CH_3)_6C_6]_2Re$, a free radical detected by its ESR spectrum. Warming this deposit back to room temperature causes dimerization with disappearance of the free radicals. Monomeric bis(hexamethylbenzene)-rhenium has a nineteen-electron configuration, which is one in excess of the favored rare-gas electronic configuration; however, its dimer of structure **XXIX** does have the eighteen-electron rare-gas configuration.

REFERENCES

1. L. F. Dahl and R. E. Rundle, *J. Chem. Phys.* **26**, 1751 (1957); M. F. Bailey and L. F. Dahl, *Inorg. Chem.* **4**, 1140 (1965); L. F. Dahl, E. Ishishi, and R. E. Rundle, *J. Chem. Phys.* **26**, 1750 (1957).
2. E. O. Brimm, M. A. Lynch, Jr., and W. J. Sesny, *J. Am. Chem. Soc.* **76**, 3831 (1954).
3. R. D. Closson, L. R. Buzbee, and G. C. Ecke, *J. Am. Chem. Soc.* **80**, 6167 (1958).
4. H. E. Podall, J. H. Dunn, and H. Shapiro, *J. Am. Chem. Soc.* **82**, 1325 (1960); F. Calderazzo, *Inorg. Chem.* **4**, 293 (1965).
5. H. E. Podall and A. P. Giraitis, *J. Org. Chem.* **26**, 2587 (1961); R. B. King, J. C. Stokes, and T. F. Korenowski, *J. Organometal. Chem.* (*Amsterdam*) **11**, 641 (1968).
6. W. Hieber and H. Fuchs, *Z. Anorg. Allgem. Chem.* **248**, 256 (1941).
7. J. C. Hileman, D. K. Huggins, and H. D. Kaesz, *Inorg. Chem.* **1**, 933 (1962).
8. A. Davison, J. A. McCleverty, and G. Wilkinson, *J. Chem. Soc.* p. 1133 (1963).
9. W. Hieber and G. Wagner, *Z. Naturforsch.* **12b**, 478 (1957); **13b**, 339 (1958).
10. R. D. Closson, J. Kozikowski, and T. H. Coffield, *J. Org. Chem.* **22**, 598 (1957).
11. W. R. McClellan, *J. Am. Chem. Soc.* **83**, 1598 (1961); H. D. Kaesz, R. B. King, and F. G. A. Stone, *Z. Naturforsch.* **15b**, 763 (1960).
12. P. M. Treichel, E. Pitcher, and F. G. A. Stone, *Inorg. Chem.* **1**, 511 (1963).
13. E. W. Abel and G. Wilkinson, *J. Chem. Soc.* p. 1501 (1959).
14. W. Hieber, R. Schuh, and H. Fuchs, *Z. Anorg. Allgem. Chem.* **248**, 243 (1941).
15. P. M. Treichel, E. Pitcher, R. B. King, and F. G. A. Stone, *J. Am. Chem. Soc.* **83**, 2593 (1961).
16. H. Wawersik and F. Basolo, *Inorg. Chem.* **6**, 1066 (1967).
17. C. G. Barraclough and J. Lewis, *J. Chem. Soc.* p. 4842 (1960).
18. D. K. Huggins, W. Fellmann, J. M. Smith, and H. D. Kaesz, *J. Am. Chem. Soc.* **86**, 4641 (1964); W. Fellmann and H. D. Kaesz, *Inorg. Nucl. Chem. Letters* **2**, 63 (1966).
19. H. D. Kaesz, W. Fellmann, G. R. Wilkes, and L. F. Dahl, *J. Am. Chem. Soc.* **87**, 2753 (1965).
20. G. Wilkinson, F. A. Cotton, and J. M. Birmingham, *J. Inorg. & Nucl. Chem.* **2**, 95 (1956).
21. M. L. H. Green, L. Pratt, and G. Wilkinson, *J. Chem. Soc.* p. 3916 (1958).
22. E. O. Fischer and M. W. Schmidt, *Angew. Chem. Intern. Ed. Engl.* **6**, 93 (1967).
23. D. K. Huggins and H. D. Kaesz, *J. Am. Chem. Soc.* **83**, 4474 (1961).
24. M. L. H. Green and G. Wilkinson, *J. Chem. Soc.* p. 4314 (1958).
25. M. Dubeck and R. A. Schell, *Inorg. Chem.* **3**, 1757 (1964).
26. R. L. Cooper, M. L. H. Green, and J. T. Moelwyn-Hughes, *J. Organometal. Chem.* (*Amsterdam*) **3**, 261 (1965).

27. R. L. Cooper and M. L. H. Green, *Z. Naturforsch.* **19b**, 652 (1964).
28. E. O. Fischer and S. Breitschaft, *Ber.* **99**, 2213 (1966).
29. E. O. Fischer and H. W. Wehner, *Ber.* **101**, 454 (1968).
30. E. O. Fischer and R. Jira, *Z. Naturforsch.* **9b**, 618 (1954); T. S. Piper, F. A. Cotton, and G. Wilkinson, *J. Inorg. & Nucl. Chem.* **1**, 165 (1955).
31. J. F. Cordes and D. Neubauer, *Z. Naturforsch.* **17b**, 791 (1962).
32. E. O. Fischer and W. Fellmann, *J. Organometal. Chem.* (*Amsterdam*) **1**, 191 (1963).
33. E. O. Fischer and K. Plesske, *Ber.* **91**, 2719 (1958).
34. M. Cais and J. Kozikowski, *J. Am. Chem. Soc.* **82**, 5667 (1960).
35. E. O. Fischer and M. Herberhold, *Experientia* Suppl. 9, 259 (1964); E. O. Fischer, H. P. Kögler, and P. Kuzel, *Ber.* **93**, 3006 (1960).
36. R. B. King and M. B. Bisnette, *Inorg. Chem.* **3**, 791 (1964).
37. T. S. Piper and G. Wilkinson, *J. Inorg. & Nucl. Chem.* **2**, 38 (1956).
38. R. B. King, *Inorg. Chem.* **6**, 30 (1967).
39. R. B. King, M. B. Bisnette, and A. Fronzaglia, *J. Organometal. Chem.* (*Amsterdam*) **5**, 341 (1966).
40. R. C. Elder, F. A. Cotton, and R. A. Schunn, *J. Am. Chem. Soc.* **89**, 3645 (1967).
41. K. K. Joshi, P. L. Pauson, A. R. Qazi, and W. H. Stubbs, *J. Organometal. Chem.* (*Amsterdam*) **1**, 471 (1964).
42. G. Winkhaus, L. Pratt, and G. Wilkinson, *J. Chem. Soc.* p. 3807 (1961).
43. G. Winkhaus, *Z. Anorg. Allgem. Chem.* **319**, 404 (1963).
44. D. Jones and G. Wilkinson, *J. Chem. Soc.* p. 2479 (1964).
45. E. O. Fischer and M. W. Schmidt, *Ber.* **99**, 2206 (1966).

QUESTIONS

1. Suggest two different routes to the "mixed" carbonyl $MnRe(CO)_{10}$, starting with $Mn_2(CO)_{10}$ and $Re_2(CO)_{10}$ as sources of manganese and rhenium, respectively.

2. Suggest specific experiments designed to determine whether the CO group lost upon decarbonylation of $CH_3COMn(CO)_5$ comes from the acyl carbonyl group or one of the terminal metal carbonyl groups.

3. Why cannot the reaction of C_2F_5I with $NaMn(CO)_5$ be used for the preparation of $C_2F_5Mn(CO)_5$?

4. Suggest preparations of the allyl derivatives $\sigma\text{-}CH_2\text{=}CHCH_2Mn(CO)_5$ and $\pi\text{-}C_3H_5Mn(CO)_4$ from $Mn_2(CO)_{10}$.

5. What are the disadvantages of preparing $C_5H_5Mn(CO)_3$ from $Mn(CO)_5Br$ and sodium cyclopentadienide over the preparations of $C_5H_5Mn(CO)_3$ discussed in the text?

CHAPTER V

Organometallic Derivatives of Iron, Ruthenium, and Osmium

Introduction

Neutral iron, ruthenium, and osmium atoms each require ten electrons from the surrounding ligands to attain the eighteen-electron configuration of the next rare gas. These ten electrons can be supplied by five monodentate ligands in a zerovalent complex. Thus, pentacarbonyliron, $Fe(CO)_5$, has the favored rare-gas configuration and is a stable compound. Alternatively, the ten electrons required to attain the rare-gas configuration can be supplied by two π-cyclopentadienyl rings. Thus, the metallocenes, $(C_5H_5)_2M$ (M = Fe, Ru, or Os), have the rare-gas configuration and, therefore, are very stable compounds. Neutral cyclopentadienylmetal (C_5H_5M) fragments of iron, ruthenium, and osmium each require five more electrons from other ligands in order to attain the eighteen-electron configuration of the next rare gas. Cyclopentadienylmetal carbonyl derivatives of iron, ruthenium, and osmium of the type $C_5H_5M(CO)_2X$ (M = Fe, Ru, or Os; X = one-electron donor) thus have the rare-gas configuration and are stable compounds.

The present availabilities of iron, ruthenium, and osmium differ greatly. Iron is the most abundant transition metal. Its organometallic chemistry has been studied in great detail. Pentacarbonyliron is inexpensive and available in tonnage quantities. It was formerly used as an antiknock agent in internal combustion engines, particularly in Europe. At the present time it is used as an intermediate for the preparation of powdered iron of controlled particle sizes for magnetic cores used in certain types of inductors and transformers. Ferrocene, $(C_5H_5)_2Fe$, is also commercially available in large quantities. Ruthenium and osmium are relatively rare and expensive. Investigations of their organometallic chemistry have been much more limited. At the present time (1968) osmium is the most expensive of the transition metals mined from natural sources. Among all of the transition metals (excluding the lanthanides and actinides), only technetium is presently more expensive than osmium.

Metallocenes

One of the most important organoiron compounds is ferrocene (or bis-cyclopentadienyliron), $(C_5H_5)_2Fe$, an orange, crystalline, diamagnetic solid, mp 173°–174°. Ferrocene is very volatile and can be purified by steam distillation or sublimation. It has a high thermal stability, decomposing only around 500°. Ferrocene is air-stable in the solid state or dissolved in organic solvents. However, in the presence of aqueous acid, ferrocene is oxidized by air or other mild oxidizing agents to give the blue ferricinium ion, $[(C_5H_5)_2Fe]^+$, which may be isolated in the form of solid salts containing large anions such as hexafluorophosphate or tetrachloroferrate(III). The ferricinium ion has a

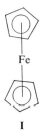

I

seventeen-electron configuration, which is one less than that of the next rare gas. For this reason, the ferricinium ion is less stable than ferrocene, and is destroyed by strong oxidizing agents such as chlorine or concentrated nitric acid. Reactions of ferrocene with strong oxidizing agents thus lead to decomposition, rather than the production, of novel π-cyclopentadienyliron derivatives.

Ferrocene was the first π-cyclopentadienyl derivative to be prepared. It was discovered by Kealy and Pauson [1] as an unexpected product of the reaction between cyclopentadienylmagnesium bromide and ferric chloride in diethyl ether solution in an unsuccessful attempt to prepare the still unknown dihydrofulvalene, C_5H_5—C_5H_5, by a coupling reaction. At about the same time, ferrocene was independently prepared by Miller, Tebboth, and Tremaine [2] in low yield by passing cyclopentadiene vapor over a special iron–molybdenum catalyst at elevated temperatures. The original discoverers of ferrocene did not recognize its unusual structure. However, shortly after its initial discovery was reported, Wilkinson, Rosenblum, Whiting, and Woodward [3] postulated the pentagonal antiprismatic structure (**I**) for ferrocene, which was subsequently confirmed by x-ray crystallography [4].

Since the original discovery of ferrocene, a variety of methods for its preparation have been developed. Some of the more interesting and useful methods for the preparation of ferrocene are summarized in the following equations:

$$2\,NaC_5H_5 + FeCl_2 \xrightarrow{\text{THF}} (C_5H_5)_2Fe + 2\,NaCl \quad [5] \qquad (a)$$

$$2\,C_5H_6 + FeO \xrightarrow{\Delta} (C_5H_5)_2Fe + H_2O \quad [6] \qquad (b)$$

The iron(II) oxide for this reaction is obtained in a finely divided state by pyrolysis of iron(II) oxalate.

$$C_5H_6 + HgCl_2 \longrightarrow C_5H_5HgCl + HCl \quad [7] \qquad (c)$$

$$2\,C_5H_5HgCl + 2\,Fe \longrightarrow (C_5H_5)_2Fe + 2\,Hg + FeCl_2$$

$$2\,C_5H_6 + FeCl_2 + 2\,Et_2NH \longrightarrow (C_5H_5)_2Fe + 2\,[Et_2NH_2]Cl \quad [5] \qquad (d)$$

The anhydrous ferrous chloride required for some of these reactions may be obtained either by heating the stoichiometric quantities of anhydrous ferric chloride and iron powder in boiling tetrahydrofuran or by heating anhydrous ferric chloride with chlorobenzene.

Ruthenocene and osmocene are also known. Pale yellow ruthenocene, $(C_5H_5)_2Ru$, mp 200°, may be obtained in $\sim 60\%$ yield by reaction of ruthenium trichloride (possibly in the presence of ruthenium metal) with sodium cyclopentadienide in tetrahydrofuran solution [8]. Colorless osmocene, $(C_5H_5)_2Os$, mp 230°, may be obtained in $\sim 23\%$ yield by reaction of osmium tetrachloride with excess sodium cyclopentadienide in 1,2-dimethoxyethane solution [9]. X-ray crystallography studies indicate ruthenocene and osmocene to have the pentagonal prismatic structure II (M = Ru or Os) rather than the pentagonal antiprismatic structure (I) of ferrocene [10].

II

The stability of the ferrocene nucleus is sufficiently high that a variety of reactions can be performed on the ferrocene system without breaking either metal–ring bond. This enables numerous substituted ferrocenes to be prepared. The chemistry of ferrocene derivatives has become so extensive that it is the subject of several review articles [11] and even one book [12].

The C_5H_5 rings in ferrocene possess an "aromaticity" similar to that of benzene. Ferrocene undergoes electrophilic substitution reactions similar to those of benzene except for the following differences: (a) ferrocene is more reactive than benzene toward electrophilic substitution, and (b) the instability of ferrocene to strong oxidizing agents makes the electrophilic nitration, sulfonation, and halogenation of ferrocene impossible, since the strongly oxidizing concentrated nitric acid, concentrated sulfuric acid, and free halogens (X_2) destroy the ferrocene system. Ferrocenyl ketones (acyl ferrocenes) can be prepared by a Friedel–Crafts reaction of ferrocene with acyl halides. Thus ferrocene reacts with acetyl chloride in the presence of aluminum chloride in dichloromethane solution to give either acetylferrocene, $CH_3COC_5H_4FeC_5H_5$, (III) or the heteroannularly disubstituted 1,1'-diacetylferrocene, $(CH_3COC_5H_4)_2Fe$ (IV), depending upon the relative quantities of reactants [13]. Similarly, formylferrocene, $C_5H_5FeC_5H_4CHO$ (V), can be prepared by electrophilic substitution of ferrocene with a mixture of dimethylformamide and phosphorus oxychloride [14]. Alkylferrocenes, such as *tert*-butylferrocene, can sometimes be prepared by a Friedel–Crafts alkylation of ferrocene with the appropriate alkyl halide in the presence of aluminum chloride [15]. Reaction of ferrocene with alkylthiolchloroformates, RSCOCl, in dichloromethane solution in the presence of an aluminum chloride catalyst gives the corresponding alkylthioferrocenyl carboxylates, $RSCOC_5H_4FeC_5H_5$ (VI: $R = CH_3$ or C_2H_5), which undergo both facile alkaline hydrolysis to salts of ferrocene carboxylic acid, and reduction with a mixture of lithium aluminum hydride and aluminum chloride to give alkylferrocenylmethyl sulfides, $RSCH_2C_5H_4FeC_5H_5$ [16].

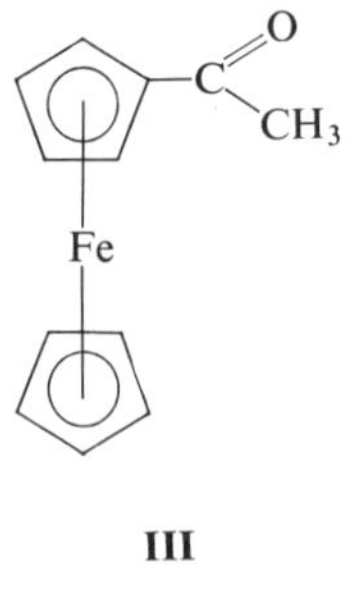

III

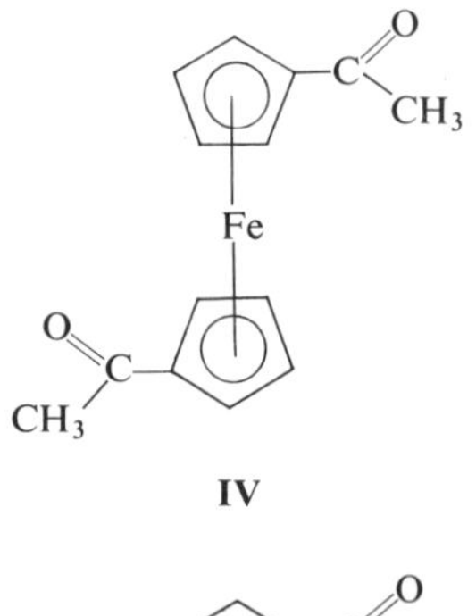

IV

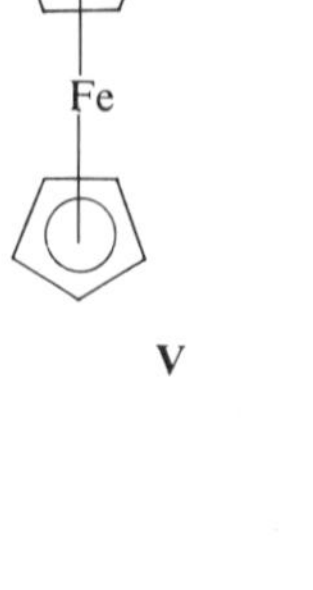

V

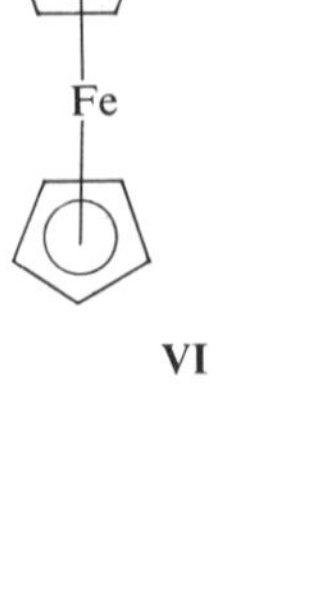

VI

Ferrocenyllithium, $C_5H_5FeC_5H_4Li$, is also useful for the preparation of numerous ferrocene derivatives. Reaction of ferrocene with *n*-butyllithium in diethyl ether solution at room temperature gives the monolithium derivative $C_5H_5FeC_5H_4Li$ [17]. Addition of *N,N,N',N'*-tetramethylethylenediamine to the diethyl ether solution containing ferrocene and *n*-butyllithium enhances the activity of the metallating agent such that good yields of ferrocenylene-1,1'-dilithium, $(LiC_5H_4)_2Fe$, are obtained [18].

Ferrocenyllithium, like other organolithium compounds, is very reactive and can be converted to numerous other ferrocene derivatives. Treatment of ferrocenyllithium with carbon dioxide followed by acidification gave ferrocenyl carboxylic acid, $C_5H_5FeC_5H_4CO_2H$ (VII) [19]; the strength of this acid is similar to that of benzoic acid. Reaction of ferrocenyllithium with an alkyl nitrate, $RONO_2$, gives the nitro derivative $C_5H_5FeC_5H_4NO_2$ (VIII), which cannot be made from ferrocene and nitric acid owing to the destructive oxidation of the ferrocene nucleus with nitric acid [20]. Ferrocenyllithium reacts with nitrous oxide, N_2O, in tetrahydrofuran solution to give azo-diferrocene, $C_5H_5FeC_5H_4N{=}NC_5H_4FeC_5H_5$ (IX), which is reduced by zinc in hydrochloric acid to give aminoferrocene, $C_5H_5FeC_5H_4NH_2$ (X) [21]. Reaction of ferrocenyllithium with tributylborate followed by hydrolysis gives ferrocenylboronic acid, $C_5H_5FeC_5H_4B(OH)_2$, which sometimes is a useful starting material for preparation of other ferrocene derivatives [22].

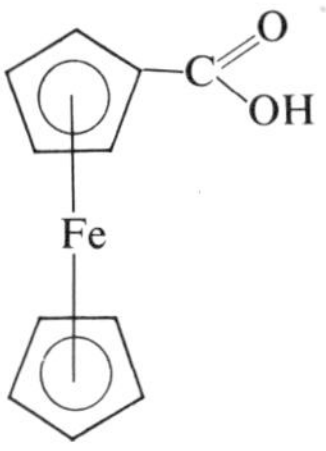

VII

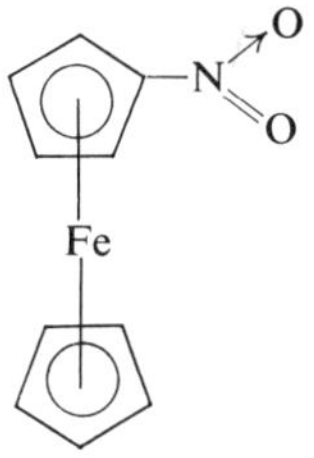

VIII

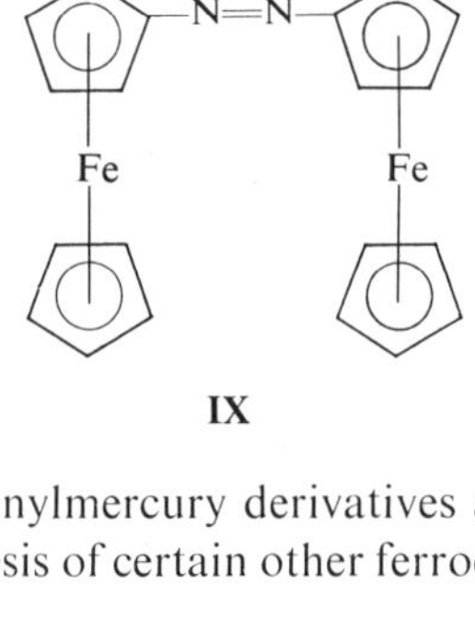

IX

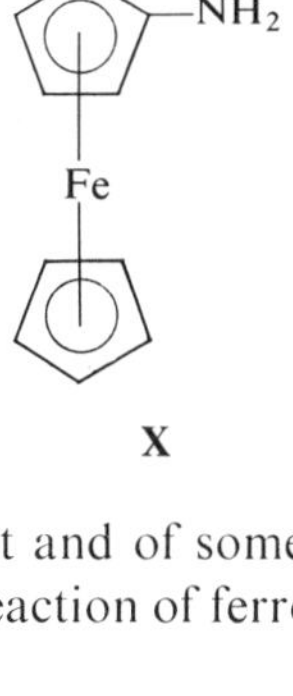

X

Ferrocenylmercury derivatives are of some interest and of some value for the synthesis of certain other ferrocene derivatives. Reaction of ferrocene with mercuric chloride in the presence of a sodium acetate buffer in a mixture of methanol and diethyl ether gives a mixture of golden yellow ferrocenyl-mercuric chloride, $C_5H_5FeC_5H_4HgCl$, mp 194°–196°, and yellow-brown 1,1'-bis(chloromercuri)ferrocene, $Fe(C_5H_4HgCl)_2$ (**XI**) [23]. Diferrocenyl-mercury, $(C_5H_5FeC_5H_4)_2Hg$, can also be prepared.

Haloferrocenes have been prepared, but by indirect methods, since treatment of ferrocene with the free halogens X_2 results in oxidation to ferricinium halides which undergo metal–ring cleavage when subjected to further action of halogen under forcing conditions. Ferrocenylmercuric halides and ferro-cenylboronic acids are useful starting materials for the preparation of halo-ferrocenes, since the bonds between ferrocenyl groups and mercury or boron are easily cleaved by appropriate halogenating agents. Thus, ferrocenyl-mercuric chloride reacts with N-halosuccinimides, $XN(CO)_2C_2H_4$, in dimethyl-formamide solution to give the haloferrocenes, $C_5H_5FeC_5H_4X$ (**XII**: X = Cl, Br, or I); similar halogenation of 1,1'-bis(chloromercuri)ferrocene, $Fe(C_5H_4HgCl)_2$ (**XI**: X = HgCl), gives the 1,1'-dihaloferrocenes, $Fe(C_5H_4X)_2$ (**XI**: X = Cl, Br, or I) [24]. Iodoferrocene, $C_5H_5FeC_5H_4I$ (**XII**: X = I), can also be made by reaction between $C_5H_5FeC_5H_4HgCl$ and iodine in dichloromethane solution [25]. Reaction of ferrocenylboronic acid, $C_5H_5FeC_5H_4B(OH)_2$, with aqueous solutions of copper(II) halides provides an alternative synthesis of the haloferrocenes, $C_5H_5FeC_5H_4X$ (**XII**: X = Cl or Br) [26].

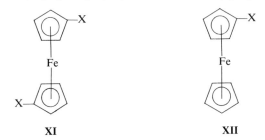

XI   XII

Haloferrocenes, like aryl and vinyl halides, are relatively unreactive. They thus form ferrocenylmagnesium halides only with difficulty upon reaction with magnesium metal in ether solvents, even in tetrahydrofuran. However, iodoferrocene, $C_5H_5FeC_5H_4I$ (**XII**: X = I), does undergo Ullman coupling with copper powder at 150°–160° to give dark orange biferrocenyl, $C_5H_5FeC_5H_4C_5H_4FeC_5H_5$ (**XIII**), mp 239°–240° [27]. Reactions of copper alkynides, $CuC\equiv CR$, with iodoferrocene in pyridine solution give the ferro-cenyl acetylenes, $C_5H_5FeC_5H_4C\equiv CR$ [28].

The π-indenyl ligand (**XIV**) is related to the π-cyclopentadienyl ligand by fusion of a benzene ring, just as naphthalene is related to benzene. Some π-indenyl derivatives of iron have been prepared which may be regarded as benzoferrocenes. Thus, reaction of ferric chloride with indenylmagnesium

XIII   XIV

bromide in diethyl ether solution gives black bisindenyliron (*sym*-dibenzo-ferrocene), $(C_9M_7)_2Fe$ **(XV)**, mp 184°–185° [*29*]. Similarly, reaction of ferrous chloride with a mixture of sodium indenide and sodium cyclopentadienide in tetrahydrofuran solution gives a ~9% yield of red-violet π-cyclopentadienyl-π-indenyliron (benzoferrocene), $C_5H_5FeC_9H_7$ **(XVI)**, mp 73°–75° [*30*]. Sub-

XV   XVI

stitution of a π-indenyl ring for a π-cyclopentadienyl ring lowers the stability of the complex but darkens its color. The two uncomplexed double bonds in the six-membered ring of the π-indenyl ligand are readily hydrogenated at atmospheric pressure in the presence of a platinum or palladium catalyst. Such hydrogenations of $(C_9H_7)_2Fe$ **(XV)** and of $C_5H_5FeC_9H_7$ **(XVI)** give the tetrahydroindenyl derivatives $(C_9H_{11})_2Fe$ **(XVII)**, mp 18°–19°, and

XVII   XVIII

$C_5H_5FeC_9H_{11}$ (**XVIII**), mp 39°–40°, respectively. Both hydrogenation products **XVII** and **XVIII** are orange, like unsubstituted ferrocene. This suggests interaction of the unsaturated six-membered ring with the electronic system of the complex in the π-indenyl derivatives **XV** and **XVI**, but no similar interaction of the saturated six-membered ring in the π-tetrahydroindenyl derivatives **XVII** and **XVIII**.

Iron Carbonyls

Iron forms the following three carbonyls: yellow liquid pentacarbonyliron, $Fe(CO)_5$, mp $-20°$, bp 101°; yellow-orange solid enneacarbonyldiiron, $Fe_2(CO)_9$, sublimes at $70°/10^{-5}$ mm; and black solid dodecacarbonyltriiron, $Fe_3(CO)_{12}$, sublimes at $70°/0.1$ mm. Pentacarbonyliron, $Fe(CO)_5$, is prepared by reaction of elemental iron with carbon monoxide at elevated temperatures and pressures [31]. It is an inexpensive commercial product and thus almost never prepared in the laboratory. Furthermore, it is the precursor to nearly all other iron carbonyl derivatives. Infrared spectra and electron diffraction data indicate structure **XIX** for $Fe(CO)_5$, with a trigonal bipyramidal arrangement of the five carbonyl groups around the central iron atom.

Pyrolysis of pentacarbonyliron at elevated temperatures results ultimately in complete decomposition, giving elemental iron and carbon monoxide. However, ultraviolet irradiation of pentacarbonyliron at or below room temperature gives enneacarbonyldiiron, $Fe_2(CO)_9$. The best results are obtained by carrying out this photolysis in acetic acid solution [32]. X-ray diffraction data indicate structure **XX** for $Fe_2(CO)_9$. Characteristic features of

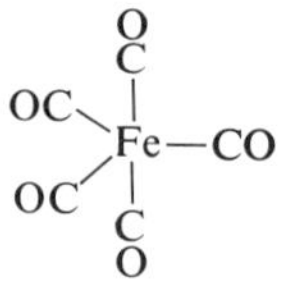

XIX 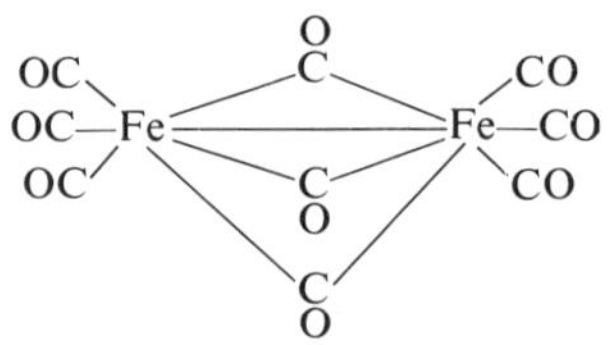   **XX**

structure are the three bridging carbonyl groups and an iron–iron bond.

Heating $Fe_2(CO)_9$ to slightly above room temperature ($\sim 50°$) causes its dissociation into $Fe(CO)_5$ and $Fe(CO)_4$ fragments according to the following equation:

$$Fe_2(CO)_9 \longrightarrow Fe(CO)_5 + Fe(CO)_4$$

The $Fe(CO)_4$ fragment is unstable and thus may react with other species present in the solution to form substituted iron carbonyl derivatives. Thus, if olefins are present, they may react with the $Fe(CO)_4$ fragment to form (olefin)$Fe(CO)_4$ complexes. This tendency to produce reactive $Fe(CO)_4$ fragments under mild conditions makes $Fe_2(CO)_9$ the most reactive of the three iron carbonyls. If $Fe_2(CO)_9$ is heated in the absence of any species which will react with the $Fe(CO)_4$ fragment produced, the $Fe(CO)_4$ fragments will trimerize, forming $Fe_3(CO)_{12}$. This was the first method by which $Fe_3(CO)_{12}$ was synthesized [33]. However, the following alternative methods for the synthesis of $Fe_3(CO)_{12}$ are much more efficient. (a) Reaction of $Fe(CO)_5$ with aqueous methanolic sodium hydroxide gives a solution containing the $HFe(CO)_4^-$ anion. Mild oxidation of this anion followed by acidification gives $Fe_3(CO)_{12}$, probably through an $HFe_3(CO)_{11}^-$ intermediate. The best oxidizing agent for this reaction appears to be freshly prepared manganese dioxide [34]. (b) Reaction of $Fe(CO)_5$ with triethylamine in aqueous solution at 80° gives the black salt $[(C_5H_5)_3NH][HFe_3(CO)_{11}]$. Acidification of this salt with sulfuric acid in methanol solution gives $Fe_3(CO)_{12}$ [35].

The structure of $Fe_3(CO)_{12}$ has been a matter of some controversy. However, this has been settled by a recent x-ray diffraction study which indicated structure **XXI** for $Fe_3(CO)_{12}$ [36]. This is similar to the structure **XX** for $Fe_2(CO)_9$, but with a bridging $Fe(CO)_4$ group replacing one of the bridging carbonyl groups.

XXI

Iron Carbonyl Anions

Several iron carbonyl anions are known. These are prepared by treatment of various iron carbonyls with various bases or reducing agents [37]. Thus, reaction of pentacarbonyliron, $Fe(CO)_5$, with aqueous or alcoholic alkali gives the colorless mononuclear ion $HFe(CO)_4^-$ with structure **XXII**. This structure may be regarded as analogous to that of $Fe(CO)_5$ (**XIX**) but with a hydride ligand (H$^-$) replacing one of the carbonyl groups. The dinegative anion $Fe(CO)_4^{2-}$ may be prepared by reduction of $Fe(CO)_5$ or $Fe_3(CO)_{12}$ with alkali metals in liquid ammonia or tetrahydrofuran. Reaction of $Fe_2(CO)_9$ with aqueous alkali gives dark red-orange $Fe_2(CO)_8^{2-}$ with structure **XXIII** containing no bridging carbonyl groups. Similar treatment of $Fe_3(CO)_{12}$ with aqueous alkali gives dark red-violet $HFe_3(CO)_{11}^-$ with structure **XXIV** similar to that of $Fe_3(CO)_{12}$, but with a bridging hydride replacing one of the bridging carbonyls. A red-black tetrairon anion $Fe_4(CO)_{13}^{2-}$ may be obtained by heating $Fe(CO)_5$ with piperidine or similar amine to 85°; this anion has structure **XXV** with one three-way bridging carbonyl group as well as three two-way bridging carbonyl groups.

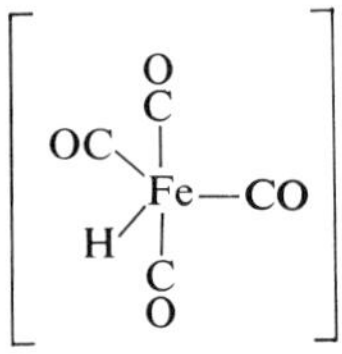

XXII

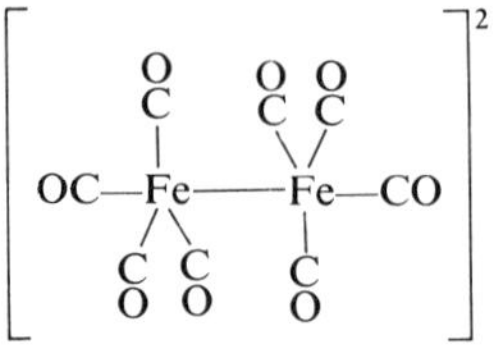

XXIII

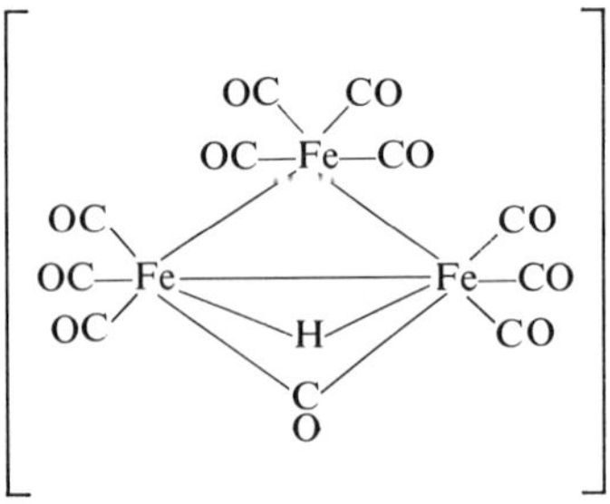

XXIV

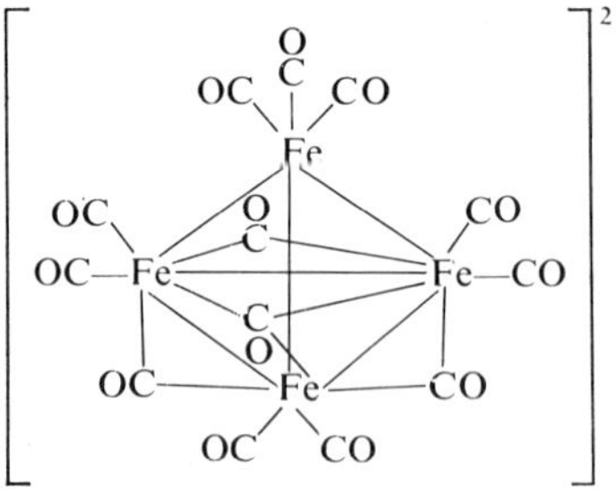

XXV

Iron Carbonyl Halide Derivatives

Several types of iron carbonyl halides have been prepared. Thus, reaction of pentacarbonyliron, $Fe(CO)_5$, with the free halogens X_2 (X = Cl, Br, or I) at low temperatures results in evolution of carbon monoxide and formation of the $Fe(CO)_4X_2$ halides according to the following equation [38]:

$$Fe(CO)_5 + X_2 \longrightarrow Fe(CO)_4X_2 + CO$$

A study of the $\nu(CO)$ frequencies in the infrared spectrum shows the $Fe(CO)_4X_2$ halides to be the *cis*-isomers, **XXVI**. Mild heating of the $Fe(CO)_4X_2$ halides results in complete loss of carbon monoxide to give the corresponding iron(II) halides. The thermal stability of the $Fe(CO)_4X_2$ halides increases in the following sequence: $Fe(CO)_4Cl_2$ (least stable) $< Fe(CO)_4Br_2 < Fe(CO)_4I_2$ (most stable). The carbonyl groups in the $Fe(CO)_4X_2$ halides can be replaced by Lewis bases such as tertiary phosphines, isocyanides, or pyridine by reaction with the ligand at room temperature.

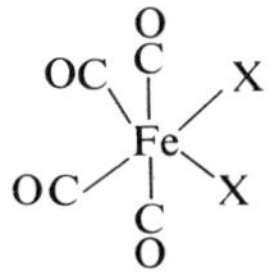

XXVI

The formation of the $Fe(CO)_4X_2$ halides from $Fe(CO)_5$ and the free halogens involves oxidation of iron(0) in the pentacoordinate $Fe(CO)_5$ to iron(II) in the hexacoordinate $Fe(CO)_4X_2$. Such reactions, which involve increases in both the formal oxidation state and coordination number, are often called *oxidative addition* reactions. Pentacarbonyliron undergoes a variety of oxidative addition reactions besides those with the halogens discussed above. Thus, reaction of $Fe(CO)_5$ with the perfluoroalkyl iodides, R_fI [$R_f = CF_3$, C_2F_5, $n\text{-}C_3F_7$, $(CF_3)_2CF$, or $n\text{-}C_7F_{15}$], at $\sim 70°$ results in the formation of the red volatile $R_fFe(CO)_4I$ derivatives according to the following oxidative addition reaction [39]:

$$Fe(CO)_5 + R_fI \longrightarrow R_fFe(CO)_4I + CO$$

The strong inductive effect of the fluorine atoms gives an R_f group an electronegativity comparable to that of a halogen atom; thus, the reactions of the perfluoroalkyl iodides, R_fI, often resemble those of the free halogens, particularly iodine.

Iodination of $Fe(CO)_5$ gives $Fe(CO)_4I_2$, as described above. Treatment of $Fe_3(CO)_{12}$ with a deficiency of iodine in boiling tetrahydrofuran proceeds somewhat differently to give $[Fe(CO)_4I]_2$ according to the following equation [40]:

$$2\ Fe_3(CO)_{12} + 3\ I_2 \longrightarrow 3\ [Fe(CO)_4I]_2$$

Side reactions and product decomposition limit the yield of $[Fe(CO)_4I]_2$ to only $\sim 2\%$. The structure of $[Fe(CO)_4I]_2$ appears to be **XXVII**, with an iron–iron bond holding the two halves together, similar to the structure of $Mn_2(CO)_{10}$ (p. 94). The compound $[Fe(CO)_4I]_2$ is unusual in being a red liquid freezing to a *white* solid at $-5°$; the reason for this color change is not understood.

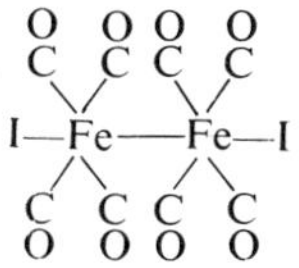

XXVII

Iron Carbonyl Sulfur Derivatives

A variety of sulfur derivatives of iron carbonyls have been prepared. Reaction of $Fe_3(CO)_{12}$ with dimethyl disulfide in boiling benzene gives red, crystalline, air-stable $[CH_3SFe(CO)_3]_2$. Chromatography of this material on an alumina column causes it to separate into two isomers [41]. The first isomer to be eluted from the column is a red solid, mp 70°–71°, which comprises ~80% of the material. The proton NMR spectrum of this isomer shows two different methyl resonances, indicating the methyl groups to be nonequivalent. This isomer is, therefore, formulated as the *anti*-isomer **XXVIII**, with one methyl group pointing up and the other methyl group pointing down. The second isomer to be eluted from the column is a red solid, mp 112°–114°, which comprises the remaining ~20% of the material. The proton NMR spectrum of this second isomer shows only one methyl resonance, indicating the methyl groups to be equivalent. This second isomer is, therefore, formulated as the *syn*-isomer **XXIX**, with both methyl groups pointing in the same direction. The reaction between $Fe_3(CO)_{12}$ and dimethyl disulfide was also found to give low yields (~1%) of a tetrametallic complex $[CH_3SFe_2(CO)_6]_2S$ indicated by x-ray crystallography [42] to have structure **XXX**.

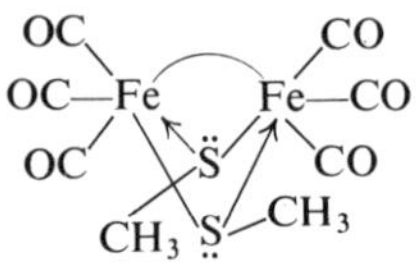

XXVIII

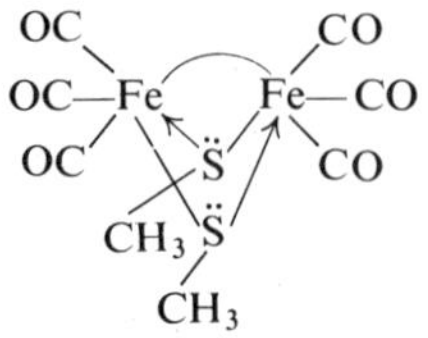

XIX

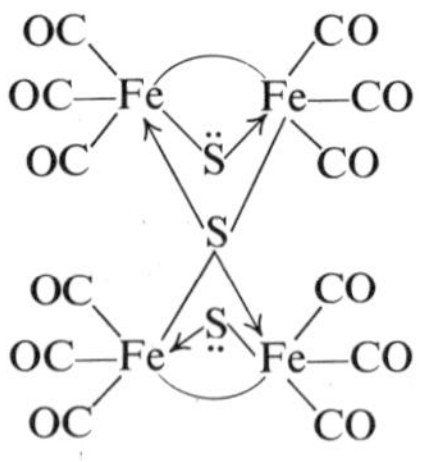

XXX

The reaction between dimethyl disulfide and $Fe(CO)_5$ does not give $[CH_3SFe(CO)_3]_2$ in an open system. However, in a closed system in the presence of carbon monoxide under pressure dimethyl disulfide and $Fe(CO)_5$ give a 60% yield of $[CH_3SFe(CO)_3]_2$. This represents the best preparation of $[CH_3SFe(CO)_3]_2$ in quantity [43]. Reaction between $Fe(CO)_5$ and dimethyl disulfide in a closed system in the *absence* of added carbon monoxide pressure gives the red insoluble coordination polymer $[(CH_3S)_2Fe(CO)_2]_n$ with repeating units of the type **XXXI** [43].

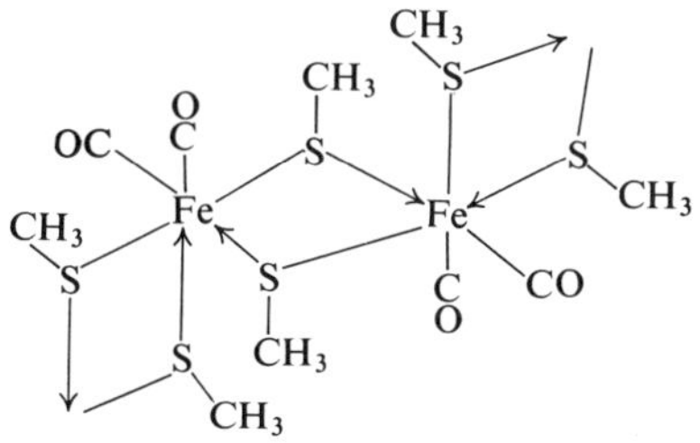

XXXI

Some iron carbonyl sulfides without organic groups attached to the sulfur can also be prepared. Thus, reaction of $Fe_3(CO)_{12}$ with episulfides gives purple volatile $Fe_3(CO)_9S_2$, mp 114°, of structure **XXXII** [44]. Reaction of the $HFe(CO)_4^-$ anion (p. 119) with aqueous sodium polysulfide followed by acidification gives a relatively low yield of red volatile $Fe_2(CO)_6S_2$, mp 46.5° [45]. X-ray crystallography of $Fe_2(CO)_6S_2$ indicates its structure to be **XXXIII** with a sulfur–sulfur bond [45].

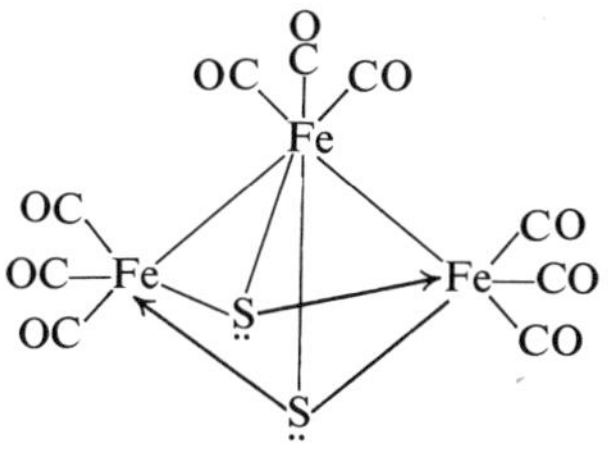

XXXII 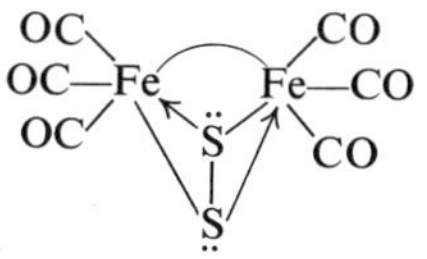   **XXXIII**

Iron Carbonyl Nitrosyl Derivatives

A variety of iron carbonyl nitrosyl derivatives are known. Many of these are prepared from the $Fe(CO)_3NO^-$ anion which can be obtained as the dark yellow potassium salt from pentacarbonyliron and potassium nitrite in methanol solution at 35°, according to the following equation [46]:

$$KNO_2 + Fe(CO)_5 \longrightarrow K[Fe(CO)_3NO] + CO + CO_2$$

Reaction of this potassium salt with mercury(II) cyanide in aqueous solution gives a red precipitate of the nonionic mercury derivative $Hg[Fe(CO)_3NO]_2$, mp 110°. This mercury derivative is insoluble in water but soluble in organic solvents. It decomposes upon standing at room temperature. Reaction of $K[Fe(CO)_3NO]$ with allyl chloride gives the red, volatile liquid π-allyl derivative $C_3H_5Fe(CO)_2NO$ (**XXXIV**) [47]. Similarly, reaction of $K[Fe(CO)_3NO]$ with the triphenylcyclopropenyl cation gives a red crystalline complex $(C_6H_5)_3C_3COFe(CO)_2NO$, formulated as **XXXV** [48].

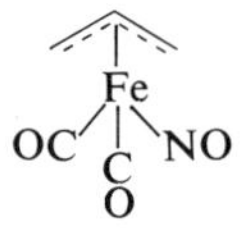

XXXIV 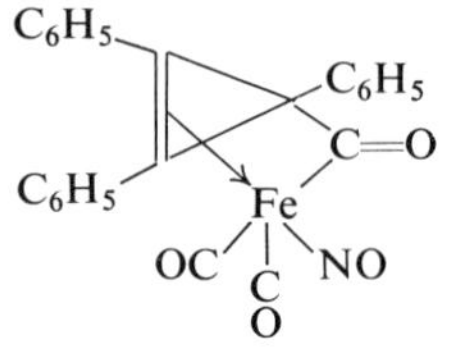   XXXV

Another iron carbonyl nitrosyl derivative of importance is red, air-sensitive $Fe(CO)_2(NO)_2$, mp 18°, bp 110°/760 mm (extrapolated), which can be prepared by the following methods.

a. Nitrosation of $HFe(CO)_4^-$ or $Fe(CO)_3NO^-$ with nitrous acid, e.g. [46]:

$$K[Fe(CO)_3NO] + NaNO_2 + 2\,CH_3COOH \longrightarrow$$
$$Fe(CO)_2(NO)_2 + NaCH_3CO_2 + KCH_3CO_2 + H_2O + CO$$

b. Reaction of $Fe_3(CO)_{12}$ with nitric oxide [49]:

$$Fe_3(CO)_{12} + 6\,NO \longrightarrow 3\,Fe(CO)_2(NO)_2 + 6\,CO$$

c. Pyrolysis of $Hg[Fe(CO)_3NO]_2$ [46]:

$$Hg[Fe(CO)_3NO]_2 \xrightarrow{\Delta} Fe(CO)_2(NO)_2 + \tfrac{1}{n}\,[HgFe(CO)_4]_n$$

This iron complex $Fe(CO)_2(NO)_2$ is a tetrahedral derivative isoelectronic with $MnCO(NO)_3$, $Co(CO)_3NO$, and $Ni(CO)_4$. Tertiary phosphines react readily with $Fe(CO)_2(NO)_2$ to replace the carbonyl groups but not the nitrosyl groups, giving compounds of the types $R_3PFe(CO)(NO)_2$ and $(R_3P)_2Fe(NO)_2$ [50].

Olefin Derivatives of Iron Carbonyls

A variety of monoolefins react with $Fe_2(CO)_9$ in a hydrocarbon solvent at 25°–50° to give yellow olefin–iron tetracarbonyl derivatives according to the following equation [51]:

$$Fe_2(CO)_9 + olefin \longrightarrow (olefin)Fe(CO)_4 + Fe(CO)_5$$

Monoolefins undergoing reactions of this type with $Fe_2(CO)_9$ include, particularly, ones with electronegative substituents such as maleic anhydride, maleimide, fumaric acid, maleic acid, acrylic acid, methyl acrylate, acrylamide, acrylonitrile, acrolein, and cinnamaldehyde. In addition, ethylene reacts similarly with $Fe_2(CO)_9$ under pressure to give yellow, volatile, liquid $C_2H_4Fe(CO)_4$. This ethylene complex and related complexes of olefins without electronegative substituents decompose to give $Fe_3(CO)_{12}$ according to the following equation:

$$3 \text{ (olefin)Fe(CO)}_4 \xrightarrow{\Delta} 3 \text{ olefin} + Fe_3(CO)_{12}$$

This reaction may proceed through unstable $Fe(CO)_4$ fragments which readily undergo trimerization (cf. the decomposition of $Fe_2(CO)_9$, p. 118). Related (olefin)$Fe(CO)_4$ complexes may be intermediates in the $Fe(CO)_5$-catalyzed isomerization of terminal to internal olefins, a reaction which may be of some industrial value [52].

A variety of conjugated diolefins react with iron carbonyls to give diene–iron tricarbonyl complexes. Butadiene thus reacts with $Fe(CO)_5$ in a bomb at ~140° to give yellow-orange butadienetricarbonyliron, $C_4H_6Fe(CO)_3$ (XXXVI), mp 19°, bp 47°–49°/0.1 mm [53]. In other cases, $Fe(CO)_5$ at ~130° or $Fe_3(CO)_{12}$ at ~80° may be used to convert conjugated diolefins into their corresponding iron tricarbonyl complexes. Conjugated diolefins known to form diene–iron tricarbonyl complexes include butadiene, isoprene, piperylene, various sorbic acid derivatives, 1,3-cyclohexadiene, octafluoro-1,3-cyclohexadiene, and 1,3-cycloheptadiene.

XXXVI

Nonconjugated diolefins readily undergo rearrangement to conjugated diolefins when heated with iron carbonyls. For this reason, diene–iron tricarbonyl complexes can rarely be prepared from nonconjugated diolefins. Thus, heating iron carbonyls with the nonconjugated 1,4-pentadiene or 1,4-cyclohexadiene derivatives gives diene–iron tricarbonyl complexes of the corresponding conjugated 1,3-pentadiene or 1,3-cyclohexadiene derivatives [54]. Boiling $Fe(CO)_5$ with 1,5-cyclooctadiene converts the 1,5-diolefin quantitatively to 1,3-cyclooctadiene, which does not form an iron tricarbonyl complex stable under the reaction conditions [55]. However, reaction of $Fe_2(CO)_9$ with 1,5-cyclooctadiene at room temperature gives yellow crystalline 1,5-cyclooctadienetricarbonyliron, $C_8H_{12}Fe(CO)_3$ (XXXVII), a rare example of an iron tricarbonyl complex of a nonconjugated diolefin. More stable iron tricarbonyl complexes of nonconjugated diolefins can be obtained from bicyclo[2,2,1]heptadiene and bicyclo[2,2,2]octadiene derivatives, where re-arrangement to a conjugated diolefin is impossible because of the inability for bridgehead carbon atoms to assume sp^2 (trigonal planar) hybridization. Thus $Fe(CO)_5$ and bicyclo[2,2,1]heptadiene (norbornadiene) react at $\sim 100°$ to produce the yellow-orange liquid complex norbornadienetricarbonyliron, $C_7H_8Fe(CO)_3$ (**XXXVIII**) [56].

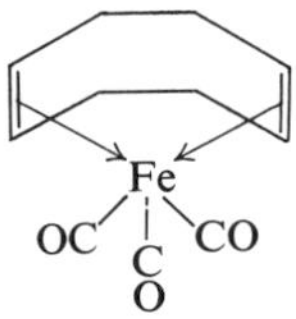

| XXXVII | XXXVIII |

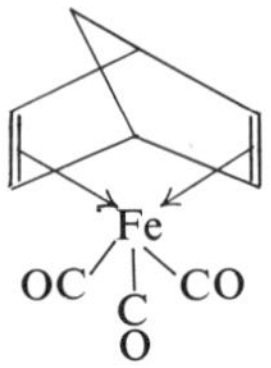

Several iron carbonyl complexes can be prepared from cycloheptatriene and related olefins. Reaction of $Fe(CO)_5$ with cycloheptatriene at 110° for prolonged periods gives a mixture of cycloheptatrienetricarbonyliron, $C_7H_8Fe(CO)_3$ (**XXXIX**), and its hydrogenation product 1,3-cycloheptadiene-tricarbonyliron, $C_7H_{10}Fe(CO)_3$ (**XL**); the latter complex can also be prepared from $Fe(CO)_5$ and 1,3-cycloheptadiene [57]. Protonation of the cyclohepta-triene complex $C_7H_8Fe(CO)_3$ (**XXXIX**) with tetrafluoroboric acid in acetic or propionic anhydride gives the yellow salt $[C_7H_9Fe(CO)_3][BF_4]$ (**XLI**) accord-ing to the following equation:

$$C_7H_8Fe(CO)_3 + HBF_4 \longrightarrow [C_7H_9Fe(CO)_3][BF_4]$$

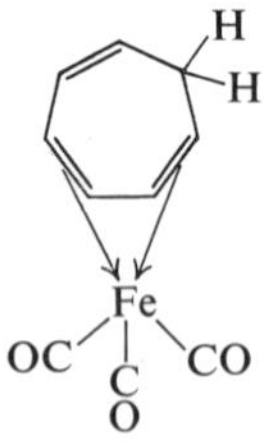

| XXXIX | XL | XLI |

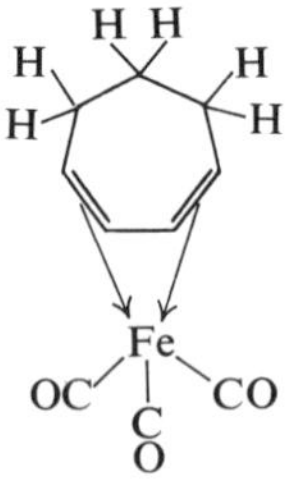

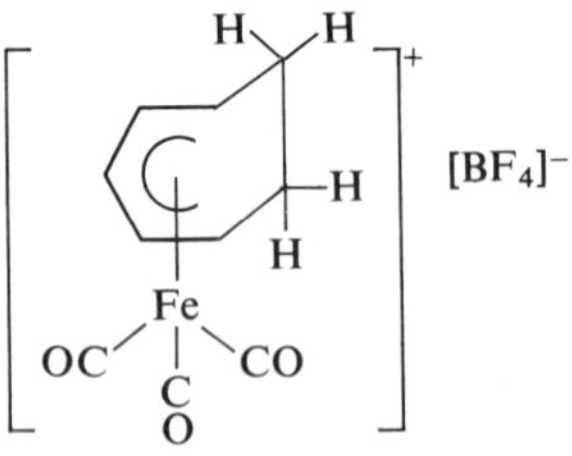

The same salt **XLI** can also be prepared by abstraction of hydride from the 1,3-cycloheptadiene complex $C_7H_{10}Fe(CO)_3$ with triphenylmethyl tetra-fluoroborate in dichloromethane solution according to the following equation:

$$C_7H_{10}Fe(CO)_3 + [(C_6H_5)_3C][BF_4] \longrightarrow [C_7H_9Fe(CO)_3][BF_4] + (C_6H_5)_3CH$$

The same cation can thus be prepared either by proton addition or hydride abstraction. The $[C_7H_9Fe(CO)_3]^+$ cation in **XLI** has five of the seven ring carbon atoms involved in the bonding to the iron atom. Nucleophilic reagents may attack either the ring or the iron atom in the cation $[C_7H_9Fe(CO)_3]^+$. Thus, potassium iodide reacts with $[C_7H_9Fe(CO)_3][BF_4]$ in acetone with gas evolution to give the maroon, crystalline, nonionic $C_7H_9Fe(CO)_2I$ (**XLII**) according to the following equation [57]:

$$[C_7H_9Fe(CO)_3][BF_4] + KI \longrightarrow C_7H_9Fe(CO)_2I + CO + KBF_4$$

In this case, the nucleophile appears to attack the iron atom. On the other hand, excess dimethylamine reacts with $[C_7H_9Fe(CO)_3][BF_4]$, attacking the ring to form the substituted 1,3-cycloheptadiene complex $[C_7H_9N(CH_3)_2]Fe(CO)_3$ according to the following equation [58]:

$$[C_7H_9Fe(CO)_3][BF_4] + 2 (CH_3)_2NH \longrightarrow$$
$$[C_7H_9N(CH_3)_2]Fe(CO)_3 + [(CH_3)_2NH_2][BF_4]$$

Azulene also forms some iron carbonyl complexes. Reaction of azulene with $Fe(CO)_5$ gives dark red $C_{10}H_8Fe_2(CO)_5$ shown by x-ray crystallography to have structure **XLIII** [59]. Other products of the reaction between azulene and iron carbonyls include $(C_{10}H_8)_2Fe_4(CO)_{10}$ and $[C_{10}H_8Fe(CO)_2]_2$.

XLII   **XLIII**

Several interesting complexes can be prepared from iron carbonyls and cyclic polyolefins with eight-membered rings such as 1,3,5-cyclooctatriene and cyclooctatetraene. Reaction of 1,3,5-cyclooctatriene with $Fe(CO)_5$ at ~130° gives yellow-orange liquid bicyclo[4,2,0]octadienetricarbonyliron, $C_8H_{10}Fe(CO)_3$ (**XLIV**), mp 8° [60]. The eight-membered ring thus becomes a bicyclic system with a six- and a four-membered ring. If the reaction between 1,3,5-cyclooctatriene and iron carbonyls is carried out under milder conditions (~80°) using the more reactive $Fe_3(CO)_{12}$, it is possible to isolate not only the bicyclic derivative **XLIV** but also the compounds $C_8H_{10}Fe(CO)_3$ (**XLV**), yellow, mp 24°, and $C_8H_{10}Fe_2(CO)_6$ (**XLVI**), red-orange, mp 101°–103°, in which the eight-membered ring is retained [61]. The complex $C_8H_{10}Fe_2(CO)_6$ (**XLVI**) is one representative of a series of red-orange complexes of the type (triene)$Fe_2(CO)_6$ formed from various cyclic triolefins (e.g., cycloheptatriene, tropone, cyclooctatriene, and cyclooctatrienone) and $Fe_2(CO)_9$ at 25°–50°.

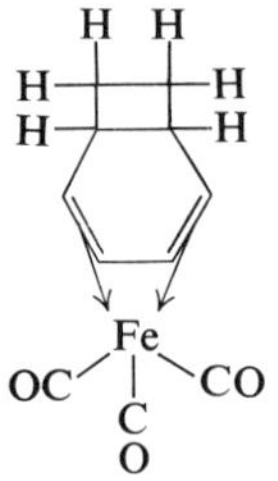

XLIV 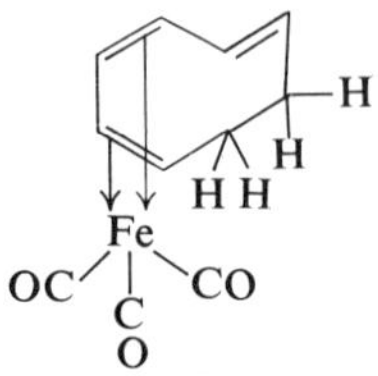   XLV 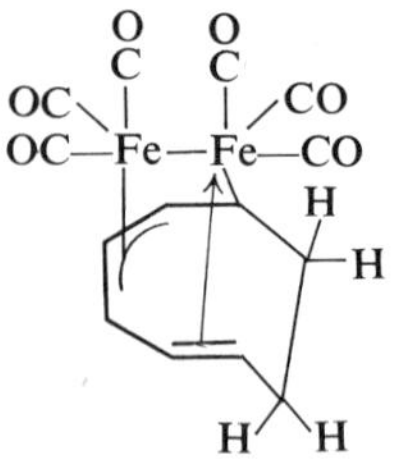   XLVI

Cyclooctatetraene forms an even greater variety of novel complexes than 1,3,5-cyclooctatriene when reacted with iron carbonyls under a variety of conditions. Reaction of cyclooctatetraene with $Fe(CO)_5$ at temperatures above ~100° gives red, air-stable, volatile cyclooctatetraenetricarbonyliron, $C_8H_8Fe(CO)_3$ (XLVII), mp 93°–95°, and lesser quantities of yellow *trans*-cyclooctatetraenebis(tricarbonyliron), $C_8H_8Fe_2(CO)_6$ (XLVIII), dec. 190° [62]. Reaction of cyclooctatetraene with $Fe_2(CO)_9$ at room temperature gives two additional $C_8H_8Fe_2(CO)_6$ isomers of structures XLIX and L; the isomer of structure XLIX is the *cis*-isomer corresponding to XLVIII. Isomers XLIX and L are both red-orange solids, mp 90°–93°. However, they exhibit a depressed mixed melting point and distinctly different proton NMR spectra. The proton NMR spectrum of XLIX exhibits two resonances at τ 5.0 and 6.0. The proton NMR spectrum of L exhibits four resonances at τ 4.2, 5.4, 5.9, and 7.5. Both of these proton NMR spectra are consistent with the number of nonequivalent protons in the structures of these complexes [63]. On the other hand, the proton NMR spectrum of $C_8H_8Fe(CO)_3$ (XLVII) in various solvents at ambient temperature exhibits a single sharp resonance at about τ 4.8. This is inconsistent with the structure XLVII supported by x-ray crystallography [64], which would be expected to have four nonequivalent pairs of protons.

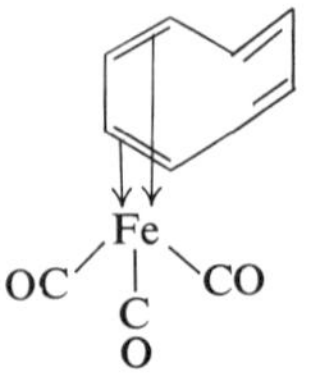

XLVII 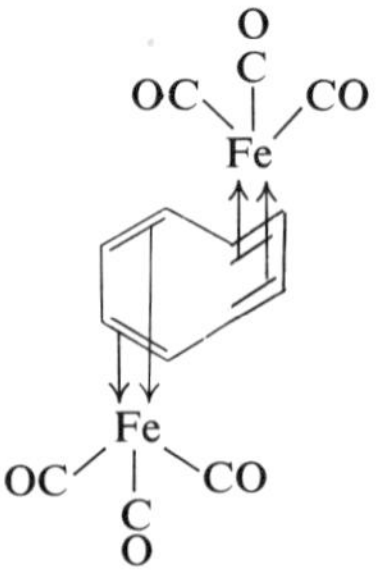   XLVIII

This apparent dilemma is resolved by postulating motion of the $Fe(CO)_3$ group about the cyclooctatetraene ring at such a rate that the eight CH groups appear equivalent on the NMR time scale, leading to only a single type of proton and hence a single resonance. This postulation was demonstrated by running the NMR spectrum of $C_8H_8Fe(CO)_3$ at temperatures below $-100°$. At these low temperatures the motion of the $Fe(CO)_3$ group about the C_8H_8 ring is retarded such that the protons of the C_8H_8 ring become nonequivalent. Upon sufficient cooling (to $\sim -150°$) the four nonequivalent protons in $C_8H_8Fe(CO)_3$ can be observed as four distinct resonances. Processes such as this motion of the $Fe(CO)_3$ group around the ring in $C_8H_8Fe(CO)_3$ are now known as *fluxional processes*; molecules exhibiting them are called *fluxional molecules*. Another fluxional molecule derived from cyclooctatetraene and iron carbonyls is black $C_8H_8Fe_2(CO)_5$ (**LI**) obtained by heating solutions of either $C_8H_8Fe_2(CO)_6$ isomer, **XLIX** or **L**; the proton NMR spectrum of $C_8H_8Fe_2(CO)_5$ (**LI**) exhibits only a single resonance at about τ 5.33 despite its apparently nonequivalent ring protons. The complex $C_8H_8Fe_2(CO)_5$ (**LI**) is unique among cyclooctatetraene–iron carbonyl complexes in containing a bridging carbonyl group.

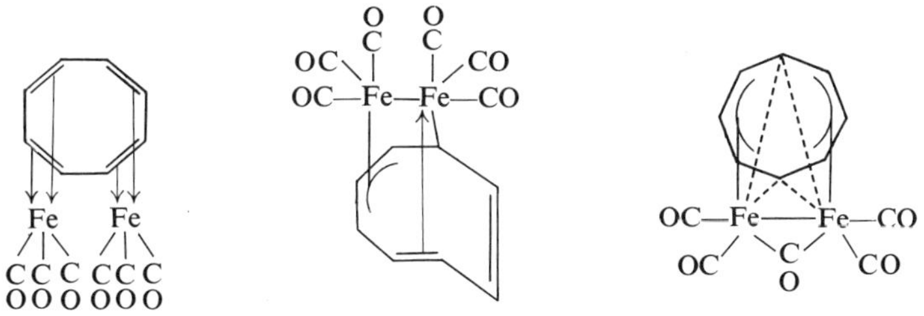

XLIX   **L**   **LI**

Cyclobutadienetricarbonyliron and Its Derivatives

An iron carbonyl complex of particular importance is yellow cyclobutadienetricarbonyliron, $C_4H_4Fe(CO)_3$ (**LII**), mp 26°, bp 68°–70°/3 mm, which is readily obtained by reaction of $Fe_2(CO)_9$ with 3,4-dichlorocyclobutene according to the following equation [65]:

$$Fe_2(CO)_9 + C_4H_4Cl_2 \longrightarrow C_4H_4Fe(CO)_3 + FeCl_2 + 6\,CO$$

Oxidation of this $C_4H_4Fe(CO)_3$ (**LII**) with ceric ammonium nitrate at 0° liberates free cyclobutadiene, which cannot be isolated, since in the absence of compounds with which it reacts this hydrocarbon rapidly dimerizes to form the tricyclooctadiene derivative **LIII**. Nevertheless, by oxidizing $C_4H_4Fe(CO)_3$ **(LII)** with ceric ion in the presence of olefinic derivatives or other compounds which react rapidly with cyclobutadiene, Pettit and co-workers have managed to develop a rather extensive chemistry of cyclobutadiene. Thus Dewar benzene (bicyclo[2,2,0]hexadiene) derivatives can be prepared by addition of cyclobutadiene to acetylenes.

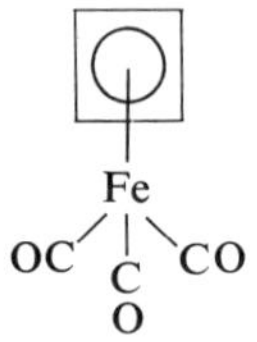

LII 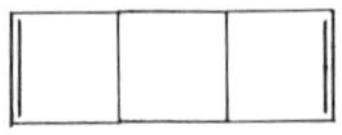   LIII

Electrophilic substitution reactions on the cyclobutadiene ring in $C_4H_4Fe(CO)_3$ **(LII)** can be carried out to produce a variety of iron tricarbonyl derivatives of substituted cyclobutadienes [66]. Friedel–Crafts acylation of $C_4H_4Fe(CO)_3$ **(LII)** with acetyl chloride in the presence of aluminum chloride gives the acetyl derivative $CH_3COC_4H_3Fe(CO)_3$ **(LIV: R = CH$_3$)**. Reaction of $C_4H_4Fe(CO)_3$ **(LII)** with a mixture of N-methylformanilide and phosphorus oxychloride gives the aldehyde $HCOC_4H_3Fe(CO)_3$ **(LIV: R = H)**. The chloromethyl derivative $ClCH_2C_4H_3Fe(CO)_3$ **(LV: X = Cl)** may be obtained by reaction between $C_4H_4Fe(CO)_3$, formaldehyde, and hydrochloric acid. Similarly, the dimethylaminomethyl derivative $(CH_3)_2NCH_2C_4H_3Fe(CO)_3$ **[LV: X = (CH$_3$)$_2$N]** may be obtained by reaction between $C_4H_4Fe(CO)_3$, formaldehyde, and dimethylamine. Mercuration of $C_4H_4Fe(CO)_3$ **(LII)** with a mixture of mercuric acetate and sodium chloride gives the chloromercuri derivative $ClHgC_4H_3Fe(CO)_3$ **(LVI)**. Further transformations of some of these substituted cyclobutadiene–iron tricarbonyl derivatives can be carried out. Thus, the ketone derivatives can be reduced to the corresponding alcohols with hydridic reducing agents.

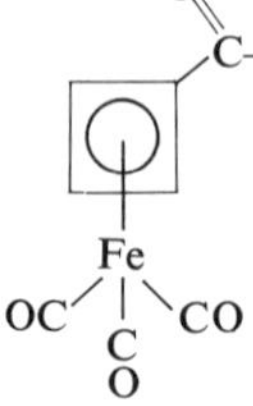

LIV 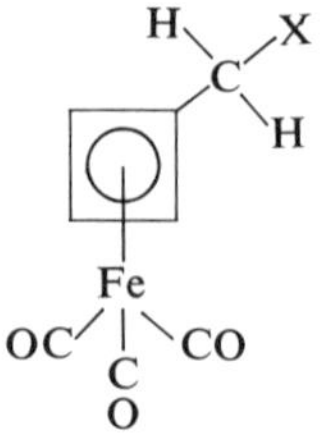   LV 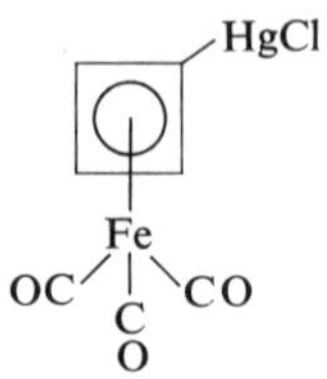   LVI

π-Allyliron Carbonyl Derivatives

Most of the known π-allyliron carbonyl derivatives have been prepared from the π-allyliron tricarbonyl halides, $C_3H_5Fe(CO)_3X$ (**LVII**: $X = Cl$, Br, or I). These may be obtained by reaction of $Fe(CO)_5$ or $Fe_2(CO)_9$ with allyl halides at temperatures below 50°, according to the following equation [67]:

$$Fe(CO)_5 + C_3H_5X \longrightarrow C_3H_5Fe(CO)_3X + 2\ CO$$

If the reactions between allyl halides and iron carbonyls are carried out at higher temperatures, ferrous halides rather than π-allyl complexes are obtained. The π-allyliron tricarbonyl halides **LVII** are yellow ($X = Cl$) to brown ($X = I$) solids subliming readily at about 40°/0.1 mm. Recent NMR evidence suggests the existence of stereoisomers of the type **LVIIa** and **LVIIb**, differing only in the orientation of the π-allyl group relative to the other ligands.

The π-allyliron tricarbonyl halides **LVII** exhibit several interesting reactions. Chromatography of the iodide $C_3H_5Fe(CO)_3I$ (**LVII**: $X = I$) on alumina in benzene solution removes the iodine atom giving $[C_3H_5Fe(CO)_3]_2$ (**LVIII**), a red pyrophoric solid subliming at 25°/1 mm [68]. In dilute solution $[C_3H_5Fe(CO)_3]_2$ dissociates readily into green $C_3H_5Fe(CO)_3$ radicals. Solutions containing these radicals react with nitric oxide to give red, air-sensitive, volatile, liquid $C_3H_5Fe(CO)_2NO$ (**XXXIV**) according to the following equation [69]:

$$C_3H_5Fe(CO)_3 + NO \longrightarrow C_3H_5Fe(CO)_2NO + CO$$

The compound $C_3H_5Fe(CO)_2NO$ may also be prepared from $KFe(CO)_3NO$ and allyl chloride.

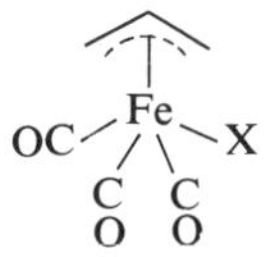

LVIIa 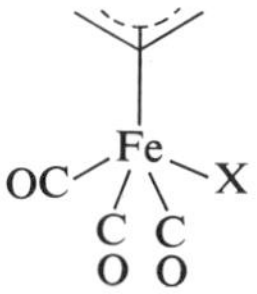   LVIIb 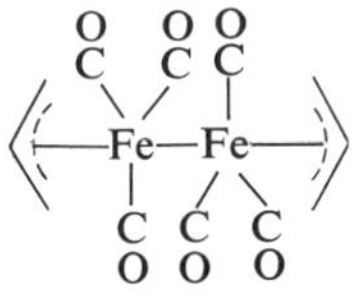   LVIII

Reactions of Iron Carbonyls with Allenes and Cumulenes; Other Delocalized Systems

Several interesting complexes can be obtained by reactions of iron carbonyls with allenes and cumulenes (polyolefins with adjacent double bonds). Reaction of allene (CH_2=C=CH_2) with $Fe_3(CO)_{12}$ in hexane solution in a closed reaction vessel gives two interesting organoiron derivatives: orange liquid $C_6H_8Fe(CO)_3$ of uncertain structure and red crystalline $[C_3H_4Fe(CO)_3]_2$, mp 88°–89°, of structure **LIX** with an unusual delocalized system of six carbon

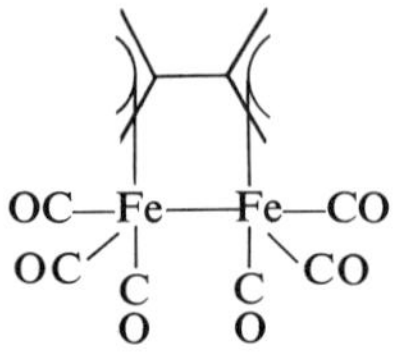

LIX

atoms bonded to two iron atoms [70]. This compound exhibits a temperature-dependent NMR spectrum apparently owing to motion of the iron tricarbonyl groups around the delocalized system. Tetraphenylallene reacts with $Fe(CO)_5$ in boiling isooctane to give red crystalline $(C_6H_5)_4C_3Fe(CO)_3$. Reaction of tetraphenylbutatriene with iron pentacarbonyl in boiling ethylcyclohexane gives red crystalline $(C_6H_5)_4C_4Fe_2(CO)_6$ [70, 71]. Unsubstituted butatriene is too unstable to be isolated and then reacted with iron carbonyls. However, generation of butatriene *in situ* from 1,4-dibromobutyne-2 and zinc in the presence of $Fe_3(CO)_{12}$ gives the unsubstituted $C_4H_4Fe_2(CO)_6$, mp 69°–70°, sublimes at 60°–80°/3 mm. X-ray crystallographic studies [72] suggest that the butatriene ligand in these (butatriene)$Fe_2(CO)_6$ complexes is bonded to each of the iron atoms by a π-allylmetal-type bond, i.e., structure **LX** for $C_4H_4Fe_2(CO)_6$.

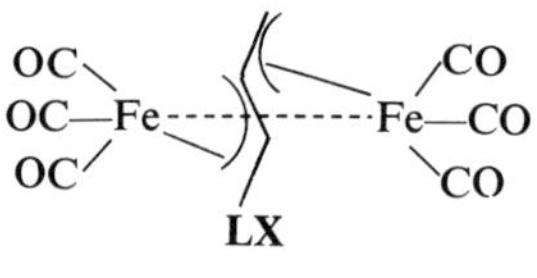

LX

Another delocalized system forming an iron carbonyl complex is trimethylenemethane [73]. Reaction of the dichloride CH_2=C(CH_2Cl)$_2$ with $Fe_2(CO)_9$

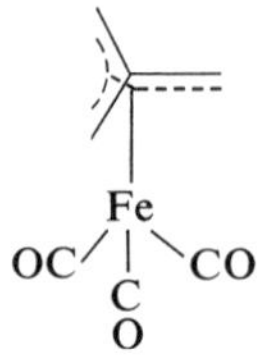

LXI

in diethyl ether solution at room temperature gives yellow trimethylene-methanetricarbonyliron, $(CH_2)_3CFe(CO)_3$, (**LXI**) mp 29°, bp 53°–55°/16 mm³. This complex **LXI** is, at present, the only known metal π-complex of the Y-shaped trimethylenemethane system. The NMR spectrum of $(CH_2)_3CFe(CO)_3$ exhibits only a single proton resonance at τ 8.00, indicating apparent equivalence of all six protons.

Reactions of Acetylenes with Iron Carbonyls

A wide variety of unusual iron carbonyl derivatives can be obtained by reactions of various acetylenes (e.g., $HC\equiv CH$ and $C_6H_5C\equiv CC_6H_5$) with iron carbonyls under a variety of conditions [74]. These include mononuclear iron tricarbonyl π-complexes of substituted cyclobutadienes (pale yellow **LXII**), cyclopentadienones (yellow **LXIII**), quinones (orange **LXIV**), and tropones (orange **LXV**), where two or three acetylene units have joined together with or without one carbon monoxide molecule to form a ring system. Under fairly mild conditions an iron tetracarbonyl complex **LXVI** (yellow) with a σ-bonded maleyl group can be obtained. Binuclear complexes (e.g., orange **LXVII** and orange **LXVIII**) with one iron tricarbonyl group in a heterocyclic ring and an iron–iron dative bond are also obtained. Other products from alkyne–iron carbonyl reactions (e.g., dark red **LXIX** and yellow **LXX**, $R = CO_2R'$) may be regarded as substituted π-cyclopentadienyl derivatives. Complex **LXXI** (black) has an iron tricarbonyl group at the bridgehead of a

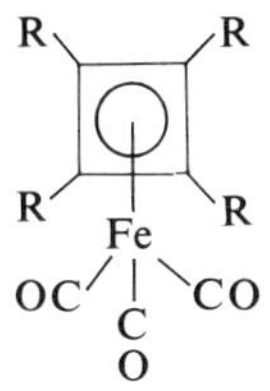

LXII 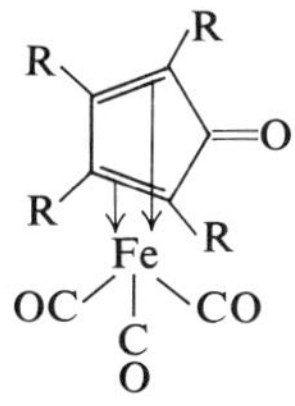   LXIII 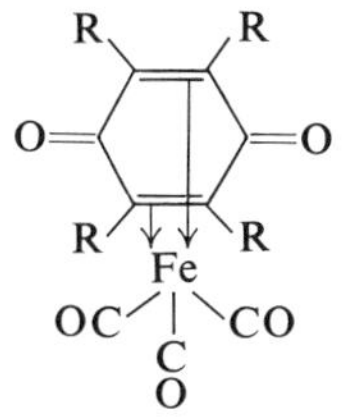   LXIV 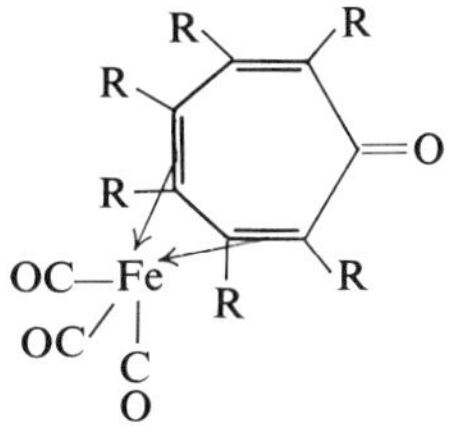   LXV

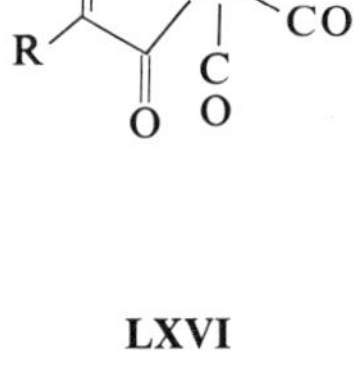

LXVI 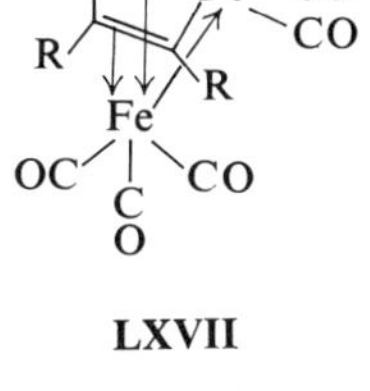   LXVII 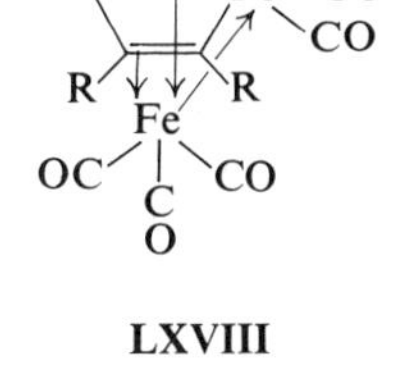   LXVIII

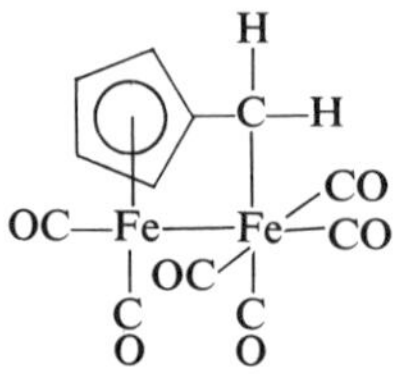

LXIX 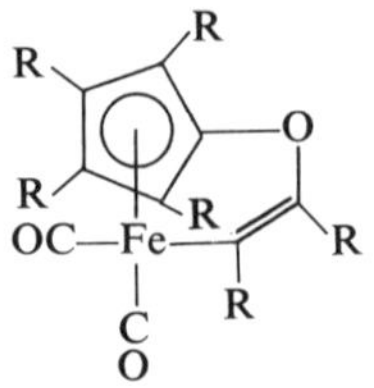 **LXX** 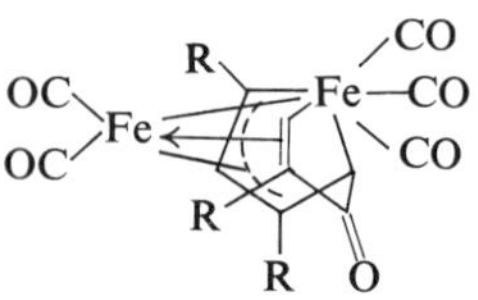 **LXXI**

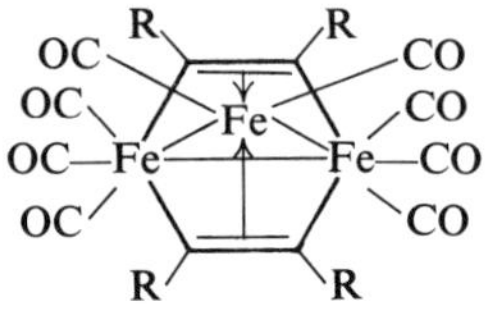

LXXII 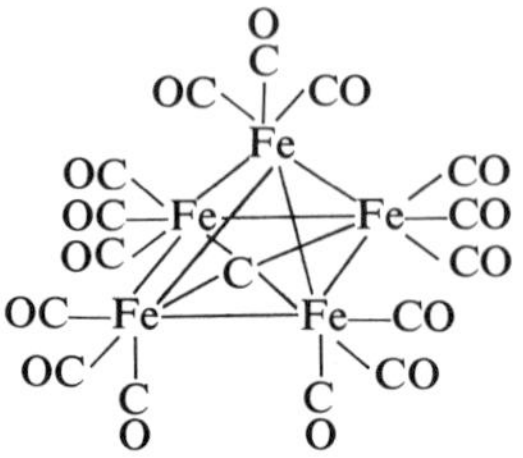 **LXXIII**

bicyclic system. Complex **LXXII** (violet) has two discrete alkyne ligands with each one bonded to all three iron atoms. Occasionally, reactions of alkynes with iron carbonyls have been found to give trace quantities ($\sim 0.1\%$ yield) of the black iron carbonyl carbide $Fe_5(CO)_{15}C$, shown by x-ray crystallography to have structure **LXXIII** with a central carbon atom within bonding distance of all five iron atoms [75].

Fluorocarbon Derivatives of Iron Carbonyls

An earlier section of this chapter (p. 121) discussed the oxidative addition reaction of $Fe(CO)_5$ with perfluoroalkyliodides, R_fI, to give the perfluoro-alkyliron iodides, $R_fFe(CO)_4I$. Several other types of fluorocarbon derivatives or iron carbonyls can be prepared. Many of their syntheses involve reactions between various fluoroolefins and iron carbonyls.

Tetrafluoroethylene, the simplest perfluoroolefin, reacts with $Fe(CO)_5$ at room temperature in the presence of ultraviolet irradiation to give $(CF_2{=}CF_2)Fe(CO)_4$ [76]. This compound is formally an (olefin)$Fe(CO)_4$ complex. However, the strong inductive effect of the four fluorine atoms withdraws much of the electron density from the π orbital of the tetrafluoro-ethylene ligand, thereby weakening the metal–olefin dative σ-bond. For this reason, a more accurate representation of the structure of the tetrafluoro-ethylene complex $(CF_2{=}CF_2)Fe(CO)_4$ would be **LXXIV**, with a three-membered ring containing one iron atom and two carbon atoms, and thus containing two iron–carbon σ-bonds. Other fluorinated olefins form similar (olefin)Fe(CO)$_4$ complexes. Reaction of iron carbonyls with tetrafluoro-ethylene under more vigorous conditions causes insertion of a second tetra-fluoroethylene unit into (CF$_2$=CF$_2$)Fe(CO)$_4$, giving white, crystalline, volatile C$_4$F$_8$Fe(CO)$_4$ (LXXV), mp 77°, with a five-membered ring containing

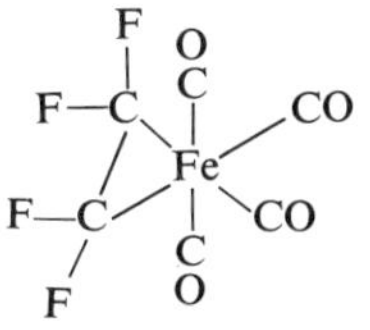

LXXIV 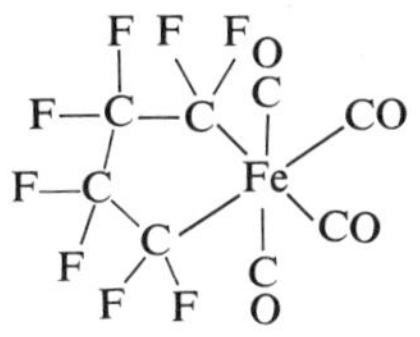 LXXV 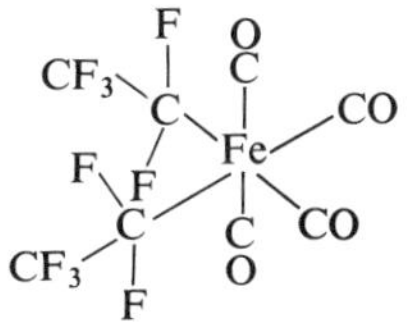 LXXVI

one iron atom and four carbon atoms [77]. A compound related to LXXV is the white, crystalline, volatile pentafluoroethyl derivative (C$_2$F$_5$)$_2$Fe(CO)$_4$ (LXXVI), mp 62°–63.5°, which can be prepared by reaction of C$_2$F$_5$COCl with Na$_2$Fe(CO)$_4$ in boiling tetrahydrofuran [78]. The Na$_2$Fe(CO)$_4$ required for this reaction is prepared from Fe$_3$(CO)$_{12}$ and excess dilute sodium amalgam in tetrahydrofuran solution. Solutions of Na$_2$Fe(CO)$_4$ in "proton-active" solvents such as water, alcohols, and ammonia cannot be used for the pre-paration of (C$_2$F$_5$)$_2$Fe(CO)$_4$, owing to the tendency for the acid chloride to react with the solvent.

Perfluorinated diolefins are relatively rare compounds. Nevertheless, reactions of iron carbonyls with both hexafluorocyclopentadiene and octa-fluoro-1,3-cyclohexadiene have been investigated. Ultraviolet irradiation of Fe(CO)$_5$ with hexafluorocyclopentadiene gives yellow C$_5$F$_6$[Fe(CO)$_4$]$_2$ (LXXVII), in which each double bond of the hexafluorocyclopentadiene is bonded to an Fe(CO)$_4$ group [79]. Reaction of Fe$_3$(CO)$_{12}$ with octafluoro-1,3-cyclohexadiene gives yellow C$_6$F$_8$Fe(CO)$_3$ (LXXVIII), mp 45°, in which both double bonds of the octafluoro-1,3-cyclohexadiene are bonded to a single Fe(CO)$_3$ group [77]. Reaction of C$_6$F$_8$Fe(CO)$_3$ with cesium fluoride in a polar solvent results in fluoride addition to give the yellow anion [C$_6$F$_9$Fe(CO)$_3$]$^-$, isolated as a cesium or tetramethylammonium salt [80].

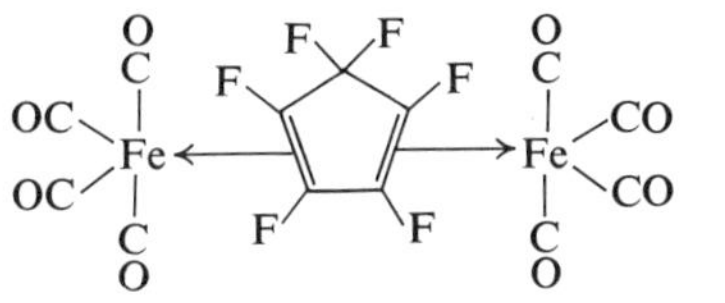

LXXVII 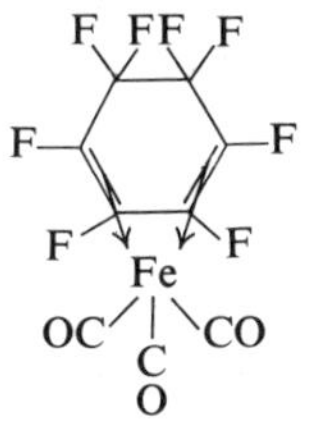 LXXVIII

Ruthenium and Osmium Carbonyl Derivatives

Ruthenium carbonyls have been known since 1936 [81]. However, the chemistry of ruthenium carbonyls has only received attention very recently, having been stimulated by the development by Bruce and Stone [82] of a relatively convenient synthesis for $Ru_3(CO)_{12}$. Their synthesis utilizes the reaction of a methanol solution of commercial hydrated ruthenium chloride with carbon monoxide at modest pressures (~ 10 atm) in the presence of zinc as a reducing agent, according to the following equation:

$$6\, RuCl_3 + 9\, Zn + 24\, CO \longrightarrow 2\, Ru_3(CO)_{12} + 9\, ZnCl_2$$

The product, $Ru_3(CO)_{12}$, is an orange, relatively air-stable solid, mp 154°–155°, sublimes at 80°–100°/0.1 mm. X-ray crystallography shows $Ru_3(CO)_{12}$ to have structure LXXIX (M = Ru) with no bridging carbonyl groups. The structure for $Ru_3(CO)_{12}$ is thus very different from that of $Fe_3(CO)_{12}$ (XXI), p. 119).

LXXIX

The mononuclear $Ru(CO)_5$, analogous to $Fe(CO)_5$, is also known. It may be prepared by heating ruthenium(III) acetylacetonate with a mixture of carbon monoxide and hydrogen under pressure [83]. Pentacarbonylruthenium, $Ru(CO)_5$, is a liquid at room temperature which decomposes on standing, evolving carbon monoxide and giving $Ru_3(CO)_{12}$. No compound $Ru_2(CO)_9$ analogous to $Fe_2(CO)_9$ is known. The compound, formerly [81] believed to be $Ru_2(CO)_9$, has now [83] been shown to be $Ru_3(CO)_{12}$.

Some tetranuclear ruthenium carbonyl hydrides have been prepared. Reaction of ruthenium trichloride with carbon monoxide in ethanol solution at atmospheric pressure gives a red solution containing unidentified ruthenium carbonyl chloride derivatives. Reaction of this red solution with carbon monoxide and hydrogen at 75°/120 atm in the presence of silver powder gives orange $[HRu(CO)_3]_4$ [84]. Reaction of $Ru_3(CO)_{12}$ with concentrated aqueous alkali followed by acidification of the resulting solution gives a mixture of $[HRu(CO)_3]_4$ and $H_2Ru_4(CO)_{13}$ [85]. Mass spectroscopy was important in establishing the formulas of these ruthenium carbonyl hydrides. These tetranuclear derivatives appear to contain tetrahedral clusters of ruthenium atoms.

A variety of reactions of $Ru_3(CO)_{12}$ have been investigated [86]. In most cases these parallel known reactions of iron carbonyls. However, the ruthenium–ruthenium bonds in $Ru_3(CO)_{12}$ are much stronger than the iron–iron bonds in $Fe_3(CO)_{12}$. As a consequence of this, the triangle of metal atoms is more often retained in reaction products of $Ru_3(CO)_{12}$ than in those of $Fe_3(CO)_{12}$. Thus, reaction of $Ru_3(CO)_{12}$ with triphenylphosphine in boiling hexane gives the dark red trinuclear derivative $[(C_6H_5)_3PRu(CO)_3]_3$; under similar conditions, $Fe_3(CO)_{12}$ and triphenylphosphine form predominantly mononuclear derivatives. Reactions of $Ru_3(CO)_{12}$ with dimethyldisulfide, 1,3-cyclohexadiene, tetramethylbiphosphine, etc., in most cases parallel analogous reactions of iron carbonyls. Cyclooctatetraene and $Ru_3(CO)_{12}$, besides giving complexes analogous to the cyclooctatetraene–iron carbonyls (p. 128), also give a complex $(C_8H_8)_2Ru_3(CO)_4$ of structure LXXX in which the triangle of ruthenium atoms is retained [87]. This compound arouses interest both by being a fluxional molecule and by containing a ruthenium atom bonded not to carbonyl groups but, instead, only to cyclooctatetraene double bonds and the other ruthenium atoms.

Osmium forms several carbonyl derivatives of interest. The mononuclear $Os(CO)_5$ may be obtained by reaction of osmium tetroxide with carbon monoxide under pressure [83]; it is an unstable, volatile liquid which is readily converted to yellow crystalline $Os_3(CO)_{12}$ (LXXIX: M = Os), with a structure analogous to $Ru_3(CO)_{12}$. The reaction between osmium tetroxide and carbon monoxide under pressure in xylene or methanol solution can be made to give good yields of $Os_3(CO)_{12}$ [88] as well as a white oxycarbonyl $Os_4O_4(CO)_{12}$ [89]. This oxycarbonyl appears to be an adduct of structure LXXXI, with OsO_4 as a Lewis acid and $Os_3(CO)_{12}$ as a Lewis base, and utilizing one of the electron pairs of the osmium–osmium bonds in the Os_3 triangle. The reaction between osmium tetroxide and carbon monoxide under pressure in the presence of additional hydrogen gives the colorless, volatile, liquid carbonyl hydride $H_2Os(CO)_4$. This carbonyl hydride is stable to $\sim100°$, in contrast to its iron and ruthenium analogues $H_2M(CO)_4$, which decompose rapidly even below room temperature. The stability of $H_2Os(CO)_4$ relative to its iron and

LXXX

LXXXI

ruthenium analogues appears to be a further consequence of the greater stability of bonds to third row $(5d)$ transition metals than corresponding bonds to first $(3d)$ and second $(4d)$ row transition metals.

Cyclopentadienylmetal Carbonyl Derivatives

The three metals iron, ruthenium, and osmium each form cyclopentadienyl-metal carbonyl derivatives of the type $[C_5H_5M(CO)_2]_2$ (M = Fe, Ru, or Os). The red-violet iron derivative $[C_5H_5Fe(CO)_2]_2$ can be prepared by heating $Fe(CO)_5$ with cyclopentadiene at 130°. This may be conveniently done by heating $Fe(CO)_5$ with excess dicyclopentadiene at this temperature [90]. Cyclopentadiene is generated under these conditions at a rate which allows it to react efficiently with $Fe(CO)_5$ as rapidly as it is formed. The yellow-orange ruthenium derivative $[C_5H_5Ru(CO)_2]_2$ is prepared by carbonylation of ruthenium trichloride to give the polymeric ruthenium carbonyl chloride $[Ru(CO)_2Cl_2]_n$ followed by reaction of this ruthenium carbonyl chloride with sodium cyclopentadienide in boiling tetrahydrofuran [91]. The yellow osmium derivative $[C_5H_5Os(CO)_2]_2$ is prepared by a method similar to the preparation of its ruthenium analogue, except that the intermediate osmium carbonyl chloride is the dinuclear derivative $[Os(CO)_3Cl_2]_2$, and that its reaction with sodium cyclopentadienide is best carried out in an autoclave at 220° using a benzene suspension [92].

Some interesting structural differences have been noted for the compounds $[C_5H_5M(CO)_2]_2$ upon descending the column of the periodic table from iron through ruthenium to osmium. In solution, these structural differences are best detected from the $\nu(CO)$ frequencies in the 2100–1700 cm^{-1} region of the infrared spectrum [92]. The iron compound $[C_5H_5Fe(CO)_2]_2$ exists primarily as the bridged isomer LXXXII (M = Fe), with barely detectable amounts of the nonbridged isomer LXXXIII (M = Fe). The ruthenium compound $[C_5H_5Ru(CO)_2]_2$ exists at room temperature as a mixture of approximately equal quantities of the bridged isomer LXXXII (M = Ru) and the nonbridged isomer LXXXIII (M = Ru). The osmium compound $[C_5H_5Os(CO)_2]_2$ exists exclusively as the nonbridged isomer LXXXIII (M = Os). Thus, as the sizes

LXXXII   LXXXIII

of the metal atoms forming the metal–metal bond increase, the tendency for bridging carbonyl groups decreases, apparently owing to the greater difficulty for the bridges to span a longer metal–metal bond.

The iron compound $[C_5H_5Fe(CO)_2]_2$ (**LXXXII**: M = Fe) can be prepared in large quantities and hence is a useful precursor to other cyclopentadienyliron carbonyl derivatives. Refluxing $[C_5H_5Fe(CO)_2]_2$ in boiling xylene for several days gives dark green air-stable $[C_5H_5FeCO]_4$, shown to have a novel structure with a tetrahedron of iron atoms, a three-way bridging carbonyl group in each face of the tetrahedron, and a cyclopentadienyl ring π-bonded to each iron atom [93]. Pyrolysis of $[C_5H_5Fe(CO)_2]_2$ under more vigorous conditions produces appreciable amounts of ferrocene.

Most of the other reactions of $[C_5H_5Fe(CO)_2]_2$ involve cleavage of the iron–iron bond with concurrent conversion of the bridging carbonyl groups to terminal carbonyl groups. Thus, treatment of $[C_5H_5Fe(CO)_2]_2$ with the halogens X_2 (X = Cl, Br, and I) gives the cyclopentadienyliron dicarbonyl halides $C_5H_5Fe(CO)_2X$ (**LXXXIV**: X = Cl, red; X = Br, dark red; X = I, brown-black) according to the following equation [94]:

$$[C_5H_5Fe(CO)_2]_2 + X_2 \longrightarrow 2\ C_5H_5Fe(CO)_2X$$

Ultraviolet irradiation of $[C_5H_5Fe(CO)_2]_2$ with dimethyl disulfide at room temperature results in similar cleavage of the iron–iron bond to give brown mononuclear $C_5H_5Fe(CO)_2SCH_3$ (**LXXXIV**: X = SCH_3) [95]. Heating this mononuclear derivative above 70° results in the elimination of 1 mole of carbon monoxide for each iron atom, giving brown-black $[C_5H_5Fe(CO)SCH_3]_2$ (**LXXXV**), which may be obtained in one step from $[C_5H_5Fe(CO)_2]_2$ by heating with dimethyl disulfide at 100° [96]. The complex $[C_5H_5Fe(CO)SCH_3]_2$ (**LXXXV**) is readily oxidized to an intense blue-green radical cation. The iron–iron bond in $[C_5H_5Fe(CO)_2]_2$ may also be cleaved with stannous chloride in boiling methanol. In this case the $SnCl_2$ group is inserted into the iron–iron bond to give orange $[C_5H_5Fe(CO)_2]_2SnCl_2$ (**LXXXVI**), mp 166°–168° [97].

LXXXIV **LXXXV** **LXXXVI**

Anionic metal carbonyl derivatives can be prepared from the $[C_5H_5M(CO)_2]_2$ compounds. Thus, reduction of $[C_5H_5Fe(CO)_2]_2$ with sodium amalgam in tetrahydrofuran solution gives an orange-brown solution containing the sodium salt $NaFe(CO)_2C_5H_5$. This sodium salt is very reactive. Upon treatment with a variety of halides it forms $RFe(CO)_2C_5H_5$ derivatives according to the following general equation [94]:

$$NaFe(CO)_2C_5H_5 + RX \longrightarrow RFe(CO)_2C_5H_5 + NaX$$

Thus, reaction of $NaFe(CO)_2C_5H_5$ with methyl iodide gives the air-sensitive, orange, waxy σ-methyl derivative $CH_3Fe(CO)_2C_5H_5$ (**LXXXVII**), mp 78°–82° [94]. Analogous ethyl, propyl, and isopropyl derivatives can also be prepared. However, reaction of $NaFe(CO)_2C_5H_5$ with *tert*-butyl chloride proceeds anomalously with isobutylene elimination to give the yellow, air-sensitive, thermally unstable (decomposing at 20°) "hydride" $HFe(CO)_2C_5H_5$ according to the following equation [98]:

$$NaFe(CO)_2C_5H_5 + (CH_3)_3CCl \longrightarrow HFe(CO)_2C_5H_5 + (CH_3)_2C{=}CH_2 + NaCl$$

The hydride $HFe(CO)_2C_5H_5$ can also be obtained by sodium borohydride reduction of $C_5H_5Fe(CO)_2Cl$. Polyfluoroaryliron derivatives can be prepared by reaction of $NaFe(CO)_2C_5H_5$ with highly fluorinated aromatic compounds [99]. Thus, $NaFe(CO)_2C_5H_5$ reacts with hexafluorobenzene to form the yellow-orange pentafluorophenyl derivative $C_6F_5Fe(CO)_2C_5H_5$ (**LXXXVIII**), mp 142°–143°, according to the following equation:

$$NaFe(CO)_2C_5H_5 + C_6F_6 \longrightarrow C_6F_5Fe(CO)_2C_5H_5 + NaF$$

LXXXVII   **LXXXVIII**

Other reactions of $NaFe(CO)_2C_5H_5$ give cyclopentadienyliron dicarbonyl derivatives with metal–metal bonds. Thus, reactions of $NaFe(CO)_2C_5H_5$ with the halides R_3ECl (E = Si, Ge, Sn, or Pb) of the heavier congeners of carbon give orange products of the type $R_3EFe(CO)_2C_5H_5$.

Iron compounds of the type $RFe(CO)_2C_5H_5$ undergo decarbonylation much less easily than corresponding molybdenum compounds of the type $RMo(CO)_3C_5H_5$. Reaction of $NaFe(CO)_2C_5H_5$ with acyl halides R'COCl gives orange acyl derivatives of the type $R'COFe(CO)_2C_5H_5$ (**LXXXIX**) [100]. Heating these acyl derivatives **LXXXIX** does not result in their decarbonylation to the corresponding $R'Fe(CO)_2C_5H_5$ derivatives; complete decomposition occurs before the decarbonylation point is reached. However, ultraviolet irradiation of $R'COFe(CO)_2C_5H_5$ derivatives in solution at room temperature gives the corresponding $R'Fe(CO)_2C_5H_5$ derivatives. Photochemical decarbonylations of $R'COFe(CO)_2C_5H_5$ derivatives provide useful preparations of the perfluoroalkyl derivatives $R_fFe(CO)_2C_5H_5$, the vinyl derivative $CH_2=CHFe(CO)_2C_5H_5$, and the phenyl derivative $C_6H_5Fe(CO)_2C_5H_5$ [101]; none of these compounds can be obtained satisfactorily from $NaFe(CO)_2C_5H_5$ and the corresponding halide [101].

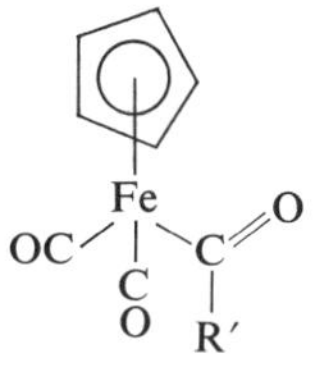

LXXXIX

A variety of reactions of $CH_3Fe(CO)_2C_5H_5$ have been investigated. Treatment of $CH_3Fe(CO)_2C_5H_5$ with carbon monoxide gives the acetyl derivative $CH_3COFe(CO)_2C_5H_5$. Similarly, reaction of $CH_3Fe(CO)_2C_5H_5$ with tertiary phosphines, R_3P, gives the substituted acetyl derivatives $CH_3COFe(CO)(PR_3)(C_5H_5)$ [102]. These reactions involve the insertion of CO into the methyl–iron bond. Treatment of $CH_3Fe(CO)_2C_5H_5$ with liquid sulfur dioxide gives the methylsulfinate $CH_3SO_2Fe(CO)_2C_5H_5$ (**XC**) [103]. In this case, SO_2 is inserted into the methyl–iron bond. Reaction of $CH_3Fe(CO)_2C_5H_5$ with trifluoroacetonitrile under pressure gives the volatile black complex $(CF_3C=NH)Fe(CO)(CF_3CN)(C_5H_5)$, apparently with structure **XCI** containing both a trifluoroacetonitrile ligand and a trifluoroacetimino group [104].

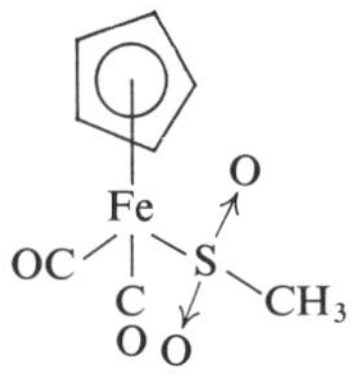

XC 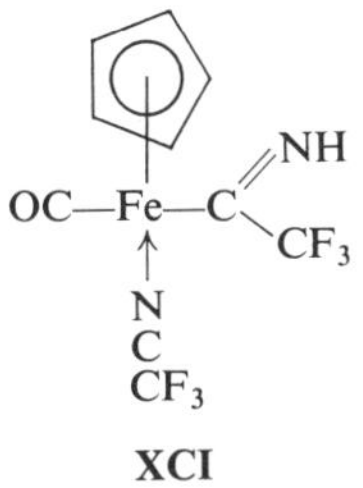   **XCI**

Some compounds of the type $RFe(CO)_2C_5H_5$ can be obtained by reactions of the halides $C_5H_5Fe(CO)_2X$ with reactive organometallic derivatives of sodium, lithium, or magnesium [94]. Thus, reaction of methylmagnesium bromide with $C_5H_5Fe(CO)_2I$ gives the same σ-methyl derivative $CH_3Fe(CO)_2C_5H_5$ that is obtained from $NaFe(CO)_2C_5H_5$ and methyl iodide. Reaction of $C_5H_5Fe(CO)_2I$ with sodium cyclopentadienide gives the orange derivative (σ-C$_5$H$_5$)Fe(CO)$_2$(π-C$_5$H$_5$), mp 46°, shown by x-ray crystallography [105] to have structure **XCII** with one π-cyclopentadienyl ring (bonded

XCII

through all five carbon atoms to the metal atom) and one σ-cyclopentadienyl ring (bonded through only one carbon atom to the metal atom). This compound **XCII** has a temperature-dependent proton NMR spectrum, indicative of a fluxional molecule [105]. Below −60° the σ-C$_5$H$_5$ ring in (σ-C$_5$H$_5$)Fe(CO)$_2$(π-C$_5$H$_5$) exhibits the expected three NMR resonances. However, at room temperature the σ-C$_5$H$_5$ ring in this complex exhibits only one NMR resonance, an indication of rapid revolution of the iron atom around the σ-C$_5$H$_5$ ring under these conditions.

Cyclopentadienyliron carbonyl cations of the type [C$_5$H$_5$Fe(CO)$_2$L]$^+$ (L = CO, olefin, R$_3$P, pyridine, etc.) have been prepared. These cations are most conveniently isolated as their hexafluorophosphates, which are precipitated upon addition of ammonium hexafluorophosphate to their aqueous solutions. The yellow ethylene cation [C$_5$H$_5$Fe(CO)$_2$C$_2$H$_4$]$^+$ may be obtained by heating a mixture of C$_5$H$_5$Fe(CO)$_2$Cl and aluminum chloride with ethylene under pressure [106], or by hydride abstraction from the ethyl derivative C$_2$H$_5$Fe(CO)$_2$C$_5$H$_5$ with triphenylmethyl tetrafluoroborate in tetrahydrofuran solution [107]. Analogous methods can be used to prepare salts of other [C$_5$H$_5$Fe(CO)$_2$(olefin)]$^+$ cations. The cyclopentadienyliron tricarbonyl cation [C$_5$H$_5$Fe(CO)$_3$]$^+$ may be prepared by reaction of a mixture of C$_5$H$_5$Fe(CO)$_2$Cl and aluminum chloride with carbon monoxide under pressure, analogous to the preparation of the ethylene cation just discussed [106]. The use of carbon monoxide under pressure for the preparation of the [C$_5$H$_5$Fe(CO)$_3$]$^+$ cation may be avoided by preparing this cation by use of the following sequence [108]:

$$NaFe(CO)_2C_5H_5 + (CH_3)_2NCOCl \xrightarrow{THF} (CH_3)_2NCOFe(CO)_2C_5H_5 + NaCl$$

$$(CH_3)_2NCOFe(CO)_2C_5H_5 + CH_3OH \longrightarrow CH_3OCOFe(CO)_2C_5H_5 + (CH_3)_2NH$$

$$CH_3OCOFe(CO)_2C_5H_5 + 2\,HCl \longrightarrow [C_5H_5Fe(CO)_3][HCl_2] + CH_3OH$$

In this reaction sequence the third CO group of the [C$_5$H$_5$Fe(CO)$_3$]$^+$ cation comes from the dimethylcarbamyl chloride.

Benzene, Cyclohexadienyl, and Cyclohexadiene Derivatives

The metals iron, ruthenium, and osmium in the 2+ formal oxidation state form $[(arene)_2M]^{2+}$ cations (M = Fe, red; M = Ru, pale yellow; M = Os, pale yellow) which are isoelectronic with the chromium, molybdenum, and tungsten derivatives $(arene)_2M$. The $[(arene)_2M]^{2+}$ cations are best obtained by heating mixtures of the metal halide, excess aluminum halide, and the free arene for several hours, followed by hydrolysis in the presence of ammonium hexafluorophosphate [109, 110]. In the cases of ruthenium and osmium, where appropriate metal halides have the metal atom in the 3+ oxidation state, aluminum powder is added to the reaction mixture as a reducing agent. The $[(arene)_2M]^{2+}$ cations increase in hydrolytic stability in going from iron through ruthenium to osmium, and by increasing the number of methyl substituents on the arene. The dibenzene–iron ion $[(C_6H_6)_2Fe]^{2+}$ is so unstable hydrolytically that it is destroyed upon hydrolysis of the reaction mixture. However, the red bis(hexamethylbenzene)–iron ion $[(CH_3)_6C_6]_2Fe^{2+}$ is so stable hydrolytically that it can be reduced in two steps, giving first the purple paramagnetic iron(I) cation $[(CH_3)_6C_6]_2Fe^+$, and then the neutral black, very unstable iron(0) derivative $[(CH_3)_6C_6]_2Fe$ [111].

Iron also forms the yellow cation $[C_5H_5FeC_6H_6]^+$ with structure **XCIII**, containing the iron atom sandwiched between the five-membered ring and the six-membered ring. Salts of this $[C_5H_5FeC_6H_6]^+$ cation were first prepared [112] by reaction of $C_5H_5Fe(CO)_2Cl$ with benzene in the presence of aluminum chloride. A more convenient preparation of the $[C_5H_5FeC_6H_6]^+$ cation (**XCIII**) utilizes the reaction between ferrocene, benzene, and aluminum chloride in the presence of aluminum powder to inhibit oxidation of the ferrocene [113]. In both cases, the reaction mixtures are hydrolyzed and $[C_5H_5FeC_6H_6][PF_6]$ isolated after addition of ammonium hexafluorophosphate.

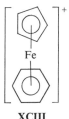

XCIII

Hydride reductions of some of these benzene–metal cations give products of interest. Lithium aluminum hydride reduction of a tetrahydrofuran solution of the $[C_5H_5FeC_6H_6]^+$ cation gives the red π-cyclohexadienyl derivative $C_5H_5FeC_6H_7$ (**XCIV**), mp 135°–136° [114]. Similar lithium aluminum hydride reduction of $[(C_6H_6)_2Ru]^+$ gives a mixture of the isomers $C_6H_6RuC_8H_8$ (**XCV**: M = Ru) and $(C_6H_7)_2Ru$ (**XCVI**) which could be identified from their NMR spectra, but which could not be separated [*114*]. An analogous iron compound $C_6H_6FeC_6H_8$ (**XCV**: M = Fe) can be obtained as orange, air-sensitive crystals, mp 102°–104°, by ultraviolet irradiation of a mixture of ferric chloride, 1,3-cyclohexadiene, and isopropylmagnesium bromide [*115*].

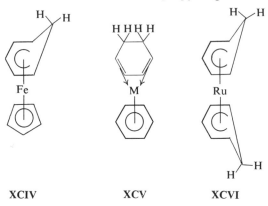

| XCIV | XCV | XCVI |

REFERENCES

1. T. J. Kealy and P. L. Pauson, *Nature* **168**, 1039 (1961).
2. S. A. Miller, J. A. Tebboth, and J. F. Tremaine, *J. Chem. Soc.* p. 632 (1952).
3. G. Wilkinson, M. Rosenblum, M. C. Whiting, and R. B. Woodward, *J. Am. Chem. Soc.* **74**, 2125 (1952).
4. J. Dunitz, L. E. Orgel, and A. Rich, *Acta Cryst.* **9**, 373 (1956).
5. G. Wilkinson, *Org. Syn.* **36**, 31 (1956).
6. R. Riemschneider and D. Helm, *Z. Naturforsch.* **14b**, 811 (1959).
7. K. Issleib and A. Brack, *Z. Naturforsch.* **11b**, 420 (1956).
8. D. E. Bublitz, W. E. McEwen, and J. Kleinberg, *Org. Syn.* **41**, 96 (1961).
9. E. O. Fischer and H. Grubert, *Ber.* **92**, 2302 (1959).
10. G. L. Hardgrove and D. H. Templeton, *Acta Cryst.* **12**, 28 (1959).
11. K. Plesske, *Angew. Chem. Intern. Ed. Engl.* **1**, 312 and 394 (1962); W. Little, *Surv. Progr. Chem.* **1**, 133 (1963); M. D. Rausch, *Can. J. Chem.* **41**, 1289 (1963).
12. M. Rosenblum, "Chemistry of Organometallic Compounds. Part I. Chemistry of the Iron Group Metallocenes: Ferrocene, Ruthenocene, Osmocene," Wiley (Interscience), New York, 1965.
13. C. R. Hauser and J. K. Lindsay, *J. Org. Chem.* **22**, 482 (1957).
14. M. Rosenblum, A. K. Banerjee, N. Danieli, R. W. Fish, and V. Schlatter, *J. Am. Chem. Soc.* **85**, 316 (1963).
15. T. Leigh, *J. Chem. Soc.* p. 3294 (1964).
16. D. E. Bublitz and G. H. Harris, *J. Organometal. Chem.* (*Amsterdam*) **4**, 404 (1965).
17. D. Mayo, P. D. Shaw, and M. D. Rausch, *Chem. & Ind.* (*London*) p. 1388 (1957).
18. M. D. Rausch and D. J. Ciappenelli, *J. Organometal. Chem.* (*Amsterdam*) **10**, 127 (1967).
19. J. Tirouflet, E. Laviron, R. Dabard, and J. Komenda, *Bull. Soc. Chim. France* p. 857 (1963).

20. H. Grubert and K. L. Rinehart, *Tetrahedron Letters* p. 16 (1959).
21. A. N. Nesmeyanov, E. G. Perevalova, and T. V. Nikitina, *Dokl. Akad. Nauk SSSR* **138**, 1118 (1961); *Tetrahedron Letters* p. 1 (1960); A. N. Nesmeyanov, T. V. Nikitina, and E. G. Perevalova, *Izv. Akad. Nauk SSSR, Ser. Khim.* p. 197 (1964).
22. H. Shechter and J. F. Helling, *J. Org. Chem.* **26**, 1034 (1961).
23. M. D. Rausch, M. Vogel, and H. Rosenberg, *J. Org. Chem.* **22**, 900 (1957).
24. R. W. Fish and M. Rosenblum, *J. Org. Chem.* **30**, 1253 (1965).
25. H. Shechter and J. F. Helling, *J. Org. Chem.* **26**, 1034 (1961).
26. A. N. Nesmeyanov, V. A. Sazonova, and V. N. Droz, *Ber.* **93**, 2717 (1960).
27. M. D. Rausch, *J. Am. Chem. Soc.* **82**, 2080 (1960).
28. M. Rosenblum, N. Brawn, J. Papenmeier, and M. Applebaum, *J. Organometal. Chem.* (*Amsterdam*) **6**, 173 (1966); M. D. Rausch, A. Siegel, and L. P. Klemann, *J. Org. Chem.* **31**, 2703 (1966); K. Schlögl and W. Steyrer, *J. Organometal. Chem.* (*Amsterdam*) **6**, 399 (1966).
29. E. O. Fischer and D. Seus, *Z. Naturforsch.* **8b**, 694 (1953).
30. R. B. King and M. B. Bisnette, *Inorg. Chem.* **3**, 796 (1964).
31. A. Mittasch, *Angew. Chem.* **30**, 827 (1928).
32. E. Speyer and H. Wolf, *Ber.* **60**, 1424 (1927).
33. H. G. Cutforth and P. W. Selwood, *J. Am. Chem. Soc.* **65**, 2414 (1943).
34. W. Hieber, *Z. Anorg. Allgem. Chem.* **204**, 165 (1932).
35. M. Heintzelor, German Patent 928,044 (1955).
36. C. H. Wei and L. F. Dahl, *J. Am. Chem. Soc.* **88**, 1821 (1966).
37. W. Hieber, W. Beck, and G. Braun, *Angew. Chem.* **22**, 795 (1960); R. B. King, *Advan. Organometal. Chem.* **2**, 157 (1964).
38. W. Hieber and G. Bader, *Ber.* **61**, 1717 (1928).
39. R. B. King, S. L. Stafford, P. M. Treichel, and F. G. A. Stone, *J. Am. Chem. Soc.* **83**, 3604 (1961).
40. F. A. Cotton and B. F. G. Johnson, *Inorg. Chem.* **6**, 2113 (1967).
41. R. B. King, *J. Am. Chem. Soc.* **84**, 2460 (1962).
42. J. M. Coleman, A. Wojcicki, P. J. Pollick, and L. F. Dahl, *Inorg. Chem.* **6**, 1236 (1967).
43. R. B. King and M. B. Bisnette, *Inorg. Chem.* **4**, 1663 (1965).
44. R. B. King, *Inorg. Chem.* **2**, 326 (1963); C. H. Wei and L. F. Dahl, *Inorg. Chem.* **4**, 493 (1965).
45. W. Hieber and J. Gruber, *Z. Anorg. Allgem. Chem.* **296**, 91 (1958); C. H. Wei and L. F. Dahl, *Inorg. Chem.* **4**, 1 (1965).
46. W. Hieber and H. Beutner, *Z. Anorg. Allgem. Chem.* **320**, 101 (1963).
47. R. Bruce, F. M. Chaudhary, G. R. Knox, and P. L. Pauson, *Z. Naturforsch.* **20b**, 73 (1965).
48. C. E. Coffey, *J. Am. Chem. Soc.* **84**, 118 (1962).
49. R. L. Mond and A. E. Wallis, *J. Chem. Soc.* **121**, 32 (1922).
50. L. Malatesta and A. Araneo, *J. Chem. Soc.* p. 3803 (1957); W. Hieber, W. Beck, and H. Tengler, *Z. Naturforsch.* **15b**, 411 (1960); D. W. McBride, S. L. Stafford, and F. G. A. Stone, *Inorg. Chem.* **1**, 386 (1962).
51. H. D. Murdoch and E. Weiss, *Helv. Chim. Acta* **46**, 1588 (1963); E. Weiss, K. Stark, J. E. Lancaster, and H. D. Murdoch, *Helv. Chim. Acta* **46**, 288 (1963).
52. T. A. Manuel, *J. Org. Chem.* **27**, 3941 (1962).
53. B. F. Hallam and P. L. Pauson, *J. Chem. Soc.* p. 642 (1958).
54. R. B. King, T. A. Manuel, and F. G. A. Stone, *J. Inorg. & Nucl. Chem.* **16**, 233 (1961).
55. J. E. Arnet and R. Pettit, *J. Am. Chem. Soc.* **83**, 2954 (1961).
56. R. Pettit, *J. Am. Chem. Soc.* **81**, 1266 (1959).

57. R. Burton, L. Pratt, and G. Wilkinson, *J. Chem. Soc.* p. 594 (1961); H. J. Dauben, Jr. and D. J. Bertelli, *J. Am. Chem. Soc.* **83**, 497 (1961).
58. F. M. Chaudhari and P. L. Pauson, *J. Organometal. Chem.* (*Amsterdam*) **5**, 73 (1966).
59. R. Burton, L. Pratt, and G. Wilkinson, *J. Chem. Soc.* p. 4290 (1960); M. R. Churchill, *Chem. Commun.* p. 450 (1966).
60. T. A. Manuel and F. G. A. Stone, *J. Am. Chem. Soc.* **82**, 366 (1960).
61. T. A. Manuel and F. G. A. Stone, *J. Am. Chem. Soc.* **82**, 6240 (1960); R. B. King, *Inorg. Chem.* **2**, 807 (1963); F. A. Cotton and W. T. Edwards, *J. Am. Chem. Soc.* **91**, 843 (1969).
62. T. A. Manuel and F. G. A. Stone, *J. Am. Chem. Soc.* **82**, 366 (1960).
63. C. E. Keller, G. F. Emerson, and R. Pettit, *J. Am. Chem. Soc.* **87**, 1388 (1965).
64. B. Dickens and W. N. Lipscomb, *J. Am. Chem. Soc.* **83**, 4862 (1961).
65. G. F. Emerson, L. Watts, and R. Pettit, *J. Am. Chem. Soc.* **87**, 131 (1965).
66. L. Watts, J. D. Fitzpatrick, and R. Pettit, *J. Am. Chem. Soc.* **87**, 3253 (1965).
67. R. A. Plowman and F. G. A. Stone, *Z. Naturforsch.* **17b**, 575 (1962); H. D. Murdoch and E. Weiss, *Helv. Chim. Acta* **45**, 1927 (1962); R. F. Heck and C. R. Boss, *J. Am. Chem. Soc.* **86**, 2580 (1964).
68. H. D. Murdoch and E. A. C. Lucken, *Helv. Chim. Acta* **47**, 1517 (1964).
69. H. D. Murdoch, *Z. Naturforsch.* **20b**, 179 (1965).
70. A. Nakamura, P. J. Kim, and N. Hagihara, *J. Organometal. Chem.* (*Amsterdam*) **3**, 7 (1965); A. Nakamura and N. Hagihara, *ibid.* p. 480.
71. A. Nakamura, P. J. Kim, and N. Hagihara, *J. Organometal. Chem.* (*Amsterdam*) **6**, 420 (1966).
72. O. S. Mills, unpublished results (1966).
73. G. F. Emerson, K. Ehrlich, W. P. Giering, and P. C. Lauterbur, *J. Am. Chem. Soc.* **88**, 3172 (1966).
74. F. L. Bowden and A. B. P. Lever, *Organometal. Chem. Rev.* **3**, 227 (1968).
75. E. H. Braye, L. F. Dahl, W. Hübel, and D. L. Wampler, *J. Am. Chem. Soc.* **84**, 4663 (1962).
76. R. Fields, M. M. Germain, R. N. Haszeldine, and P. W. Wiggans, *Chem. Commun.* p. 243 (1967).
77. H. H. Hoehn, L. Pratt, K. F. Watterson, and G. Wilkinson, *J. Chem. Soc.* p. 2738 (1961).
78. R. B. King, S. L. Stafford, P. M. Treichel, and F. G. A. Stone, *J. Am. Chem. Soc.* **83**, 3604 (1961).
79. R. E. Banks, T. Harrison, R. N. Haszeldine, A. B. P. Lever, T. F. Smith, and J. B. Walton, *Chem. Commun.* p. 30 (1965).
80. G. W. Parshall and G. Wilkinson, *J. Chem. Soc.* p. 1132 (1962).
81. W. Manchot and W. J. Manchot, *Z. Anorg. Allgem. Chem.* **226**, 385 (1936).
82. M. I. Bruce and F. G. A. Stone, *Chem. Commun.* p. 684 (1966).
83. F. Calderazzo and F. L'Eplattenier, *Inorg. Chem.* **6**, 1220 (1967).
84. J. W. S. Jamieson, J. V. Kingston, and G. Wilkinson, *Chem. Commun.* p. 569 (1966).
85. B. F. G. Johnson, R. D. Johnston, J. Lewis, and B. H. Robinson, *Chem. Commun.* p. 851 (1966).
86. J. P. Candlin, K. K. Joshi, and D. T. Thompson, *Chem. & Ind.* (*London*) p. 1960 (1966); B. F. G. Johnson, R. D. Johnston, P. L. Josty, J. Lewis, and I. G. Williams, *Nature* **213**, 901 (1967).
87. M. J. Bennett, F. A. Cotton, and P. Legzdins, *J. Am. Chem. Soc.* **89**, 6797 (1967).
88. C. W. Bradford and R. S. Nyholm, *Chem. Commun.* p. 384 (1967).
89. B. F. G. Johnson, J. Lewis, I. G. Williams, and J. Wilson, *Chem. Commun.* p. 391 (1966).

90. T. S. Piper, F. A. Cotton, and G. Wilkinson, *J. Inorg. & Nucl. Chem.* **1**, 165 (1955); R. B. King and F. G. A. Stone, *Inorg. Syn.* **7**, 110 (1963).
91. E. O. Fischer and A. Vogler, *Z. Naturforsch.* **17b**, 421 (1962).
92. E. O. Fischer, A. Vogler, and K. Noack, *J. Organometal. Chem. (Amsterdam)* **7**, 135 (1967).
93. R. B. King, *Inorg. Chem.* **5**, 2227 (1966).
94. T. S. Piper and G. Wilkinson, *J. Inorg. & Nucl. Chem.* **3**, 104 (1956).
95. R. B. King and M. B. Bisnette, *Inorg. Chem.* **4**, 482 (1965).
96. R. B. King, P. M. Treichel, and F. G. A. Stone, *J. Am. Chem. Soc.* **83**, 3600 (1961).
97. F. Bonati and G. Wilkinson, *J. Chem. Soc.* p. 179 (1964).
98. M. L. H. Green and P. L. I. Nagy, *J. Organometal. Chem. (Amsterdam)* **1**, 58 (1963).
99. R. B. King and M. B. Bisnette, *J. Organometal. Chem. (Amsterdam)* **2**, 38 (1964).
100. R. B. King, *J. Am. Chem. Soc.* **85**, 1918 (1963).
101. R. B. King and M. B. Bisnette, *J. Organometal. Chem. (Amsterdam)* **2**, 15 (1964).
102. J. P. Bibler and A. Wojcicki, *Inorg. Chem.* **5**, 889 (1966).
103. J. P. Bibler and A. Wojcicki, *J. Am. Chem. Soc.* **88**, 4862 (1966).
104. R. B. King and K. H. Pannell, *J. Am. Chem. Soc.* **90**, 3984 (1968).
105. M. J. Bennett, Jr., F. A. Cotton, A. Davison, J. W. Faller, S. J. Lippard, and S. M. Morehouse, *J. Am. Chem. Soc.* **88**, 4371 (1966).
106. E. O. Fischer and K. Fichtel, *Ber.* **94**, 1200 (1961).
107. M. L. H. Green and P. L. I. Nagy, *J. Organometal. Chem. (Amsterdam)* **1**, 58 (1963).
108. R. B. King, M. B. Bisnette, and A. Fronzaglia, *J. Organometal. Chem. (Amsterdam)* **5**, 341 (1966).
109. E. O. Fischer and R. Böttcher, *Ber.* **89**, 2397 (1956).
110. E. O. Fischer and R. Böttcher, *Z. Anorg. Allgem. Chem.* **291**, 305 (1957).
111. E. O. Fischer and F. Röhrscheid, *Z. Naturforsch.* **17b**, 483 (1962).
112. M. L. H. Green, L. Pratt, and G. Wilkinson, *J. Chem. Soc.* p. 989 (1960).
113. A. N. Nesmeyanov, N. A. Vol'kenau, and I. N. Bolesova, *Tetrahedron Letters* p. 1725 (1963); R. B. King, *Organometal. Syn.* **1**, 138 (1965).
114. G. Winkhaus, L. Pratt, and G. Wilkinson, *J. Chem. Soc.* p. 3807 (1961); D. Jones, L. Pratt, and G. Wilkinson, *ibid.* p. 4458 (1962).
115. E. O. Fischer and J. Müller, *Z. Naturforsch.* **17b**, 776 (1962).

SUPPLEMENTARY READING

1. M. Rosenblum, "Chemistry of the Iron Group Metallocenes: Ferrocene, Ruthenocene, Osmocene." Wiley (Interscience), New York, 1965.
2. R. Pettit and G. F. Emerson, Diene-iron carbonyl complexes and related species. *Advan. Organometal. Chem.* **1**, 1 (1964).
3. M. I. Bruce and F. G. A. Stone, Dodecacarbonyltriruthenium. *Angew. Chem. Intern. Ed. Engl.* **7**, 427 (1968).

QUESTIONS

1. Account for the differences in the structures of $[Fe(CO)_4I]_2$ and $[Mn(CO)_4I]_2$.
2. Why do the tertiary phosphines and other Lewis base ligands react with $Fe(CO)_4I_2$ at *room temperature* to replace carbonyl groups, whereas similar Lewis base ligands react similarly with the similarly octahedral $Cr(CO)_6$ only at elevated temperatures ($>100°$)?
3. Outline two different ways for converting $Fe(CO)_5$ to the π-allyl derivative $C_3H_5Fe(CO)_2NO$.
4. Suggest reasons for the different structures for $Fe_3(CO)_{12}$ and $Ru_3(CO)_{12}$.
5. Predict the NMR spectra of the two isomers $C_6H_6RuC_6H_8$ and $(C_6H_7)_2Ru$.

CHAPTER VI

Organometallic Derivatives of Cobalt, Rhodium, and Iridium

Introduction

Neutral cobalt, rhodium, and iridium each require nine electrons from the surrounding ligands in order to attain the favored eighteen-electron configuration of the next rare gas. Carbonyl derivatives of the types $RM(CO)_4$, $TM(CO)_3$, and $C_5H_5M(CO)_2$ (R = one-electron donor; T = three-electron donor; M = Co, Rh, or Ir) and their substitution products have the favored eighteen-electron rare-gas configuration. A variety of compounds of these types have been prepared. In addition, numerous square planar rhodium(I) and iridium(I) derivatives with only a sixteen-electron configuration are known. These often undergo oxidative addition reactions to give octahedral rhodium(III) and iridium(III) derivatives which have the eighteen-electron configuration of the next rare gas. Organocobalt compounds also include a variety of octahedral cobalt(III) derivatives and a few four-coordinate cobalt(II) derivatives with cobalt–carbon σ-bonds. The most stable π-cyclopentadienyl derivatives of cobalt, rhodium, and iridium are the biscyclopentadienylmetal(III) cations $[(C_5H_5)_2M]^+$ which have the favored eighteen-electron rare-gas configuration.

Cobalt Carbonyls

Cobalt forms three isoleptic ("pure") cobalt carbonyls: orange $Co_2(CO)_8$, black $Co_4(CO)_{12}$, and black $Co_6(CO)_{16}$. All are solids which are fairly air-sensitive. Octacarbonyldicobalt, $Co_2(CO)_8$, is best prepared by treatment of a cobalt salt of a weak acid (e.g., acetate or carbonate) with a mixture of carbon monoxide and hydrogen at elevated temperatures (160°) and pressures (200 atm) according to the following equation [1]:

$$2\ CoX_2 + 8\ CO + 2\ H_2 \longrightarrow Co_2(CO)_8 + 4\ HX$$

Warming $Co_2(CO)_8$ to slightly above room temperature releases some carbon monoxide, giving $Co_4(CO)_{12}$ according to the following equation:

$$2\,Co_2(CO)_8 \longrightarrow Co_4(CO)_{12} + 4\,CO$$

Unfortunately it is difficult to carry out the decomposition of $Co_2(CO)_8$ to give a good yield of $Co_4(CO)_{12}$, since side reactions giving metallic cobalt appear to occur concurrently. Furthermore it is difficult to separate $Co_4(CO)_{12}$ efficiently from these by-products. A more efficient preparation of $Co_4(CO)_{12}$ utilizes the hydrogen reduction of a hydrocarbon-soluble cobalt salt of a weak acid (e.g., cobalt 2-ethylhexanoate) in the presence of a heptane solution of $Co_2(CO)_8$ according to the following equation [2]:

$$3\,Co_2(CO)_8 + 2\,CoX_2 + 2\,H_2 \longrightarrow 2\,Co_4(CO)_{12} + 4\,HX$$

Reduction of $Co_4(CO)_{12}$ with potassium metal in tetrahydrofuran solution gives the hexacobalt salt $K_4Co_6(CO)_{14}$. Oxidation of this salt with aqueous ferric chloride gives the neutral $Co_6(CO)_{16}$ [3].

These cobalt carbonyls have interesting structures. X-ray crystallography [4] indicates crystalline $Co_2(CO)_8$ to have structure **I** with a cobalt–cobalt bond and two bridging carbonyl groups. However a study of the temperature-dependence of the $\nu(CO)$ frequencies of solutions of $Co_2(CO)_8$ indicates the presence of a second isomer, **II**, with the two $Co(CO)_4$ halves held together solely by a cobalt–cobalt bond [5]. X-ray crystallography [6] indicates crystal-line $Co_4(CO)_{12}$ to have structure **III** with a tetrahedral cluster of cobalt atoms

I   II   III

and three of the twelve carbonyl groups bridging pairs of basal cobalt atoms. The crystal structure of $Co_6(CO)_{16}$ has not yet been determined, since it was discovered so recently. However, the $\nu(CO)$ infrared frequencies in $Co_6(CO)_{16}$ resemble closely those of $Rh_6(CO)_{16}$, suggesting similar structures for the two derivatives. The structure of $Rh_6(CO)_{16}$ has been found by x-ray crystallo-graphy [7] to contain an octahedral cluster of metal atoms, two terminal carbonyl groups bonded to each metal atom and three-way bridging groups in alternate faces of the octahedron.

Other Cobalt Carbonyl Derivatives

An important preparative reagent in cobalt carbonyl chemistry is the tetracarbonylcobalt$(-I)$ anion, $Co(CO)_4^-$, which may be prepared from $Co_2(CO)_8$ by the following four general methods:

$$Co_2(CO)_8 + 2\,Na \longrightarrow 2\,NaCo(CO)_4 \qquad [8] \qquad\qquad\text{(a)}$$

$$3\,Co_2(CO)_8 + 12\,\text{base} \longrightarrow 2\,[Co^{II}(\text{base})_6][Co(CO)_4]_2 + 8\,CO \qquad [9] \qquad\text{(b)}$$

In this reaction one-third of the cobalt is converted to cobalt(II) and the other two-thirds of the cobalt is converted to cobalt$(-I)$. Suitable bases for reaction (b) are relatively strong σ-donors but weak π-acceptors such as alcohols, ethers, ketones, and amines.

$$Co_2(CO)_8 + 2\,R_3P \longrightarrow [(R_3P)_2Co^I(CO)_3][Co(CO)_4] + CO \qquad [10] \qquad\text{(c)}$$

$$Co_2(CO)_8 + 5\,RNC \longrightarrow [Co^I(RNC)_5][Co(CO)_4] + 4\,CO \qquad [11] \qquad\text{(d)}$$

In reactions (c) and (d) half of the cobalt is converted to cobalt(I) and the remaining half of the cobalt is converted to cobalt$(-I)$.

An aqueous ammoniacal solution of cobalt(II) can be converted to a solution of $Co(CO)_4^-$ by treatment with carbon monoxide at atmospheric pressure in the presence of an appropriate reducing agent such as potassium cyanide or sodium dithionite [12].

The $Co(CO)_4^-$ anion may be used to prepare a variety of $RCo(CO)_4$ derivatives. Many of the $RCo(CO)_4$ compounds are unstable, possibly owing to the reactivity of the five-coordinate central cobalt atom. Acidification of $Co(CO)_4^-$ with a strong nonoxidizing acid (e.g., H_2SO_4) gives the hydride $HCo(CO)_4$ as a toxic, malodorous gas decomposing rapidly in the condensed phase at room temperature [13]. Similar treatment of $Co(CO)_4^-$ with methyl iodide gives the light yellow, very unstable σ-methyl derivative $CH_3Co(CO)_4$, mp $-44°$, decomposing at $> -35°$ [14]. Reaction of $Co(CO)_4^-$ with acetyl chloride gives the unstable acetyl derivative $CH_3COCo(CO)_4$ [15]. Similar treatment of $Co(CO)_4^-$ with heptafluorobutyryl chloride, C_3F_7COCl, at room temperature gives, after distillation, the yellow liquid heptafluoropropyl derivative $C_3F_7Co(CO)_4$, bp $44°/16$ mm [16]. The presumed intermediate $C_3F_7COCo(CO)_4$ undergoes decarbonylation so rapidly that it is not obtained. The greater stability of $C_3F_7Co(CO)_4$ relative to $CH_3Co(CO)_4$ is typical for fluorocarbon derivatives and arises from partial cobalt–carbon(C_3F_7) multiple bonding, enhanced by the electronegative fluorine atoms. Compounds with a $Co(CO)_4$ group bonded to silicon or its heavier congeners are more stable than analogous compounds with a $Co(CO)_4$ group bonded to carbon. Thus, reaction between $Co(CO)_4^-$ and silyl iodide, SiH_3I gives the yellow liquid silyl derivative $H_3SiCo(CO)_4$, mp $-53°$, bp $112°$, which is much more stable than its methyl analogue $CH_3Co(CO)_4$ [17]. Substituted silylcobalt tetra-carbonyls may often be prepared from $Co_2(CO)_8$ and silanes according to the following equation [18]:

$$2\ R_3SiH + Co_2(CO)_8 \longrightarrow 2\ R_3SiCo(CO)_4 + H_2$$

where $R = CH_3$, C_6H_5, Cl, etc. Analogous derivatives with cobalt–tin bonds may be prepared by reaction of $Co(CO)_4^-$ with trialkyltin halides, e.g. [19]:

$$Co(CO)_4^- + (CH_3)_3SnCl \longrightarrow (CH_3)_3SnCo(CO)_4 + Cl^-$$
$$\text{(pale yellow solid,}$$
$$\text{sublimes at } 40°/0.1 \text{ mm,}$$
$$\text{mp } 74°)$$

Compounds of the type $TCo(CO)_3$ (T = three-electron donor) have the favored eighteen-electron rare-gas configuration. These include the π-allylic cobalt tricarbonyls. The first such compound to be prepared was the π-1-methylallyl derivative $CH_3C_3H_4Co(CO)_3$, obtained as a brown-red liquid, mp $-60°$, bp $35°/1$ mm, by reaction of $HCo(CO)_4$ with butadiene [20]. This compound $CH_3C_3H_4Co(CO)_3$ was found by NMR studies to exist as a mixture of stereoisomers IV where R = H; R′ = CH_3 and IV where R = CH_3; R′ = H, differing in the orientation of the methyl substituent on the π-allylic ligand relative to the metal atom. The unsubstituted π-allyltricarbonylcobalt, $C_3H_5Co(CO)_3$ (IV: R = R′ = H), is obtained as a red-yellow liquid, mp $-33°$, by reaction of allyl bromide with $NaCo(CO)_4$ in diethyl ether solution at room temperature [21]. An unusual π-allylic cobalt tricarbonyl is the red air-sensitive π-cycloheptatrienyl derivative $C_7H_7Co(CO)_3$ (V) obtained in low yield by ultraviolet irradiation of a mixture of cycloheptatriene and $Co_2(CO)_8$ [22]. The NMR spectrum of $C_7H_7Co(CO)_3$ (V) exhibited a single sharp resonance at room temperature, indicating a fluxional system with rapid motion of the $Co(CO)_3$ group around the seven-membered ring.

IV   V

Another compound of the type $TCo(CO)_3$ is nitrosyltricarbonylcobalt, $Co(CO)_3NO$, a dark-red liquid, mp $-1°$, bp $50°/760$ mm. This may be prepared by either of the following two methods.

a. Nitrosation of $Co(CO)_4^-$, e.g. [23]:

$$Co(CO)_4^- + HNO_2 \xrightarrow{\text{water}} Co(CO)_3NO + CO + OH^-$$

The nitrous acid required to effect this reaction may be generated from nitrite and a weak acid (acetic acid or carbon dioxide) or from nitric oxide and water.

   b. Reaction of $Co_2(CO)_8$ with nitric oxide [24]:

$$Co_2(CO)_8 + 2\ NO \longrightarrow 2\ Co(CO)_3NO + 2\ CO$$

Reaction of $Co(CO)_3NO$ with tertiary phosphines replaces one or two carbonyl groups but not the nitrosyl group, giving compounds of the types $R_3PCo(CO)_2NO$ and $(R_3P)_2Co(CO)(NO)$ [25].

   Further types of cobalt carbonyl derivatives of interest are the purple, relatively stable, volatile methinyltricobalt enneacarbonyls $YCCo_3(CO)_9$, which have structure **VI** containing a tetrahedral cluster of one carbon atom and three cobalt atoms as YC and $Co(CO)_3$ groups, respectively. This structure **VI** for the $YCCo_3(CO)_9$ compounds is formally related to that of $Co_4(CO)_{12}$ by replacing one $Co(CO)_3$ group with a YC group. Most of the $YCCo_3(CO)_9$ derivatives can be prepared by reaction of $Co_2(CO)_8$ with the YCX_3 trihalides according to the following equation [26]:

$$9\ Co_2(CO)_8 + 4\ YCX_3 \longrightarrow 4\ YCCo_3(CO)_9 + 36\ CO + 6\ CoX_2$$

where $Y = F$, Cl, Br, I, CH_3, C_6H_5, H, CF_3, CH_3OCO, etc.; $X = Cl$, Br, or I. (If Y is also a halogen, X cannot be lighter than Y.) In addition, the $YCCo_3(CO)_9$ compounds of the type $RCH_2CCo_3(CO)_9$ (**VI**: $Y = RCH_2$) may be prepared by reaction of the alkyne–dicobalt hexacarbonyls $(RC_2H)Co_2(CO)_6$ with strong mineral acids [27].

VI

   Several interesting types of compounds can be prepared from reactions between various alkynes and cobalt carbonyls. A large variety of alkynes ($YC{\equiv}CY$) react readily with $Co_2(CO)_8$ at room temperature in inert hydrocarbon solvents to give the red alkyne–dicobalt hexacarbonyls $(Y_2C_2)Co_2(CO)_6$ ($Y = H$, CH_3, C_6H_5, CF_3, CO_2CH_3, CN, etc.) which have structure **VII** containing a tetrahedral cluster of two carbon atoms and two cobalt atoms as YC and $Co(CO)_3$ groups, respectively [28]. This structure **VII** for the $(Y_2C_2)Co_2(CO)_6$ compounds is formally related to that of $Co_4(CO)_{12}$ by replacing two $Co(CO)_3$ groups with YC groups, or to that of $YCCo_3(CO)_9$ by replacing one $Co(CO)_3$ group with a YC group. Reaction of the alkyne–

dicobalt hexacarbonyls $(Y_2C_2)Co_2(CO)_6$ (VII: $Y = H$, CH_3, C_6H_5, etc.) with excess carbon monoxide at $70°-75°/210$ atm gives the red complexes $[Y_2C_2(CO)_2]Co_2(CO)_7$ with structure VIII similar to the bridged form of $Co_2(CO)_8$ (I) but with a bridging lactone carbene replacing one of the bridging carbonyl groups [29]. Reaction of the alkyne–dicobalt hexacarbonyls $(Y_2C_2)Co_2(CO)_6$ [VII: $Y = H$, CH_3, $(CH_3)_3C$, C_6H_5, etc.] with an excess of a hindered alkyne $YC{\equiv}CY$ ($Y = $ *tert*-butyl, etc.) gives purple $(Y_2C_2)_3Co_2(CO)_4$ compounds of structure IX (the Y groups may be different) with a six-carbon bridge forming both π-allylic–cobalt and cobalt–carbon σ-bonds [30].

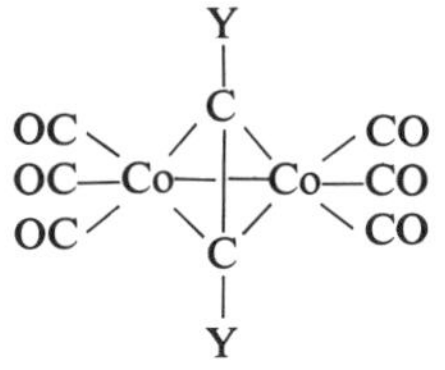

VII 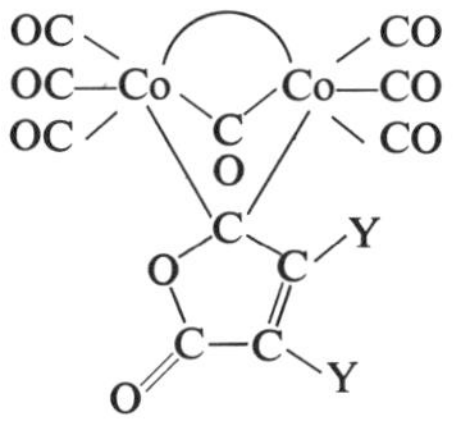   VIII 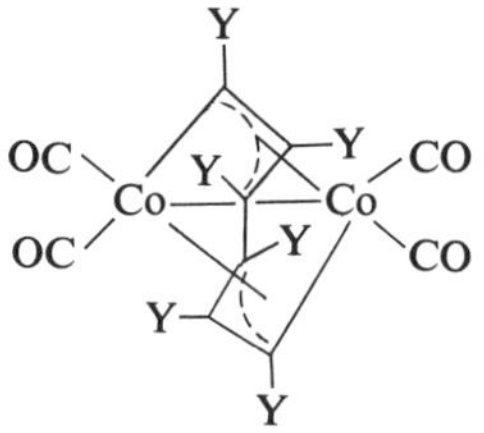   IX

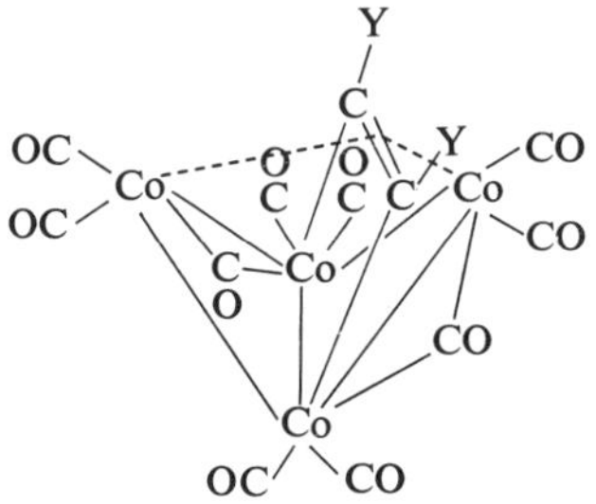

X

Degradation of these $(Y_2C_2)_3Co_2(CO)_4$ compounds (IX) gives the arenes C_6Y_6 by joining the two ends of the six-carbon chain. Alkynes react with $Co_4(CO)_{12}$ (III) to give the dark-blue complexes $Y_2C_2Co_4(CO)_{10}$ of structure X [31]. In this reaction, the closed tetrahedral cluster of cobalt atoms in $Co_4(CO)_{12}$ (III) has become the more open cluster of cobalt atoms in $(Y_2C_2)Co_4(CO)_{10}$ (X) [32].

Rhodium and Iridium Carbonyls

Rhodium forms the two carbonyls $Rh_4(CO)_{12}$ and $Rh_6(CO)_{16}$ which may be obtained by the reductive carbonylation of rhodium trichloride under various conditions [33, 34]. The structures of these carbonyls have been found by x-ray crystallography to be similar to those of their cobalt analogues (p. 149) [6, 35]. A $Rh_2(CO)_8$ analogous to $Co_2(CO)_8$ has been reported [33] but has not yet been studied by modern crystallographic and spectroscopic techniques.

A rhodium carbonyl derivative which has been studied in much greater detail is the red volatile rhodium carbonyl chloride $[Rh(CO)_2Cl]_2$, mp 124°–125°. This may be conveniently prepared by reaction of rhodium trichloride with carbon monoxide at atmospheric pressure [33, 36]. Reactions of $[Rh(CO)_2Cl]_2$ with various neutral and charged ligands produce various square planar rhodium(I) carbonyl derivatives, e.g.:

$$[Rh(CO)_2Cl]_2 + 4 R_3P \longrightarrow \underset{\text{(yellow)}}{2 (R_3P)_2Rh(CO)Cl} + 2 CO \quad [37] \qquad \text{(a)}$$

$$[Rh(CO)_2Cl]_2 + 2 Cl^- \longrightarrow \underset{\substack{\text{[pale yellow, isolated} \\ \text{as } (C_4H_9)_4N^+ \text{ salt]}}}{2 [Rh(CO)_2Cl_2]^-} \quad [38] \qquad \text{(b)}$$

$$\underset{\substack{2 \text{ RCOCH}_2\text{COR} \\ (\beta\text{-diketone})}}{[Rh(CO)_2Cl]_2 +} \xrightarrow{\text{K}_2\text{CO}_3} \underset{\text{(red volatile)}}{2 (RCOCHCOR)R(CO)_2} + HCl \quad [39] \qquad \text{(c)}$$

In addition, $[Rh(CO)_2Cl]_2$ reacts with silver carboxylates to give the red-green rhodium carbonyl carboxylates $[Rh(CO)_2CO_2R]_2$ ($R = CH_3$, etc.) [40].

Reaction of a methanol solution of iridium trichloride with carbon monoxide at elevated temperatures and pressures gives yellow $Ir_4(CO)_{12}$. X-ray crystallographic studies [41] demonstrate $Ir_4(CO)_{12}$ to have structure XI with a tetrahedral cluster of the four iridium atoms. However, no bridging carbonyl groups are present in $Ir_4(CO)_{12}$, indicating that its structure is *not* analogous to that of $Co_4(CO)_{12}$ (III). A compound $Ir_2(CO)_8$ analogous to $Co_2(CO)_8$ has been reported [42] but has not yet been investigated by modern crystallographic and spectroscopic techniques.

XI   XII

Certain square planar sixteen-electron iridium(I) carbonyl chlorides have received attention because of their tendency to undergo oxidative addition reactions to give octahedral eighteen-electron iridium(III) carbonyl chloride derivatives. Carbonylation of iridium trichloride at elevated temperatures gives brown volatile $Ir(CO)_3Cl$ [43]. The yellow triphenylphosphine derivative trans-$[(C_6H_5)_3P]_2Ir(CO)Cl$ (XII), "Vaska's catalyst," has received even greater attention. It is most conveniently prepared by heating a mixture of hydrated iridium trichloride and excess triphenylphosphine in a good "carbonylating" solvent such as 2-methoxyethanol or dimethylformamide [44]. The following equations exemplify oxidative addition reactions of trans-$[(C_6H_5)_3P]_2Ir(CO)Cl$ (XII).

$$[(C_6H_5)_3P]_2Ir(CO)Cl + H_2 \rightleftharpoons [(C_6H_5)_3P]_2IrH_2(CO)Cl \quad [45] \qquad (a)$$
(pale yellow)

$$[(C_6H_5)_3P]_2Ir(CO)Cl + O_2 \rightleftharpoons [(C_6H_5)_3P]_2IrO_2(CO)Cl \quad [46] \qquad (b)$$
(orange)

$$[(C_6H_5)_3P]_2Ir(CO)Cl + HgCl_2 \longrightarrow [(C_6H_5)_3P]_2Ir(HgCl)(CO)Cl_2 \quad [47] \qquad (c)$$
(white)

$$[(C_6H_5)_3P]_2Ir(CO)Cl + CF_2{=}CF_2 \rightleftharpoons [(C_6H_5)_3P]_2Ir(C_2F_4)(CO)Cl \quad [48] \qquad (d)$$
(white)

The reactions of XII with hydrogen, oxygen, and tetrafluoroethylene are reversible, whereas those of XII with more vigorous reagents such as mercuric chloride are irreversible.

π-Cyclopentadienyl Derivatives

Several biscyclopentadienylcobalt derivatives are known. Reaction of cobalt(II) chloride with sodium cyclopentadienide in tetrahydrofuran solution gives violet-black biscyclopentadienylcobalt (cobaltocene), $(C_5H_5)_2Co$, mp $173°–174°$ [49]. Cobaltocene has a nineteen-electron configuration (one in excess of the next rare gas) and hence exhibits the expected paramagnetism (1.76 BM) for a complex with one unpaired electron. Oxidation of $(C_5H_5)_2Co$ with air in aqueous solution proceeds according to the following equation [50]:

$$4\,(C_5H_5)_2Co + O_2 + 4\,H^+ \longrightarrow (C_5H_5)_2Co^+ + 2\,H_2O$$

The eighteen-electron biscyclopentadienylcobalt(III) (cobalticinium) ion may be isolated as its yellow PF_6^- salt by adding ammonium hexafluorophosphate to its aqueous solution. The ion $(C_5H_5)_2Co^+$ is stable to strong oxidizing agents. However, reactive lithium and sodium derivatives react with $(C_5H_5)_2Co^+$ to give compounds containing a diene-bonded substituted π-cyclopentadiene ligand as well as a π-cyclopentadienyl ligand. Thus, reduction of $(C_5H_5)_2Co^+$ with lithium aluminum hydride gives red π-cyclo-pentadienyl-π-cyclopentadienecobalt, $C_5H_5CoC_5H_6$ (**XIII**: M = Co), mp 98°–99° [51]. Similarly, sodium cyclopentadienide reacts with $(C_5H_5)_2Co^+$ in tetrahydrofuran to give the red-brown bimetallic complex $(C_5H_5)_2CoC_5H_4Co(C_5H_5)_2$ (**XIV**) [52].

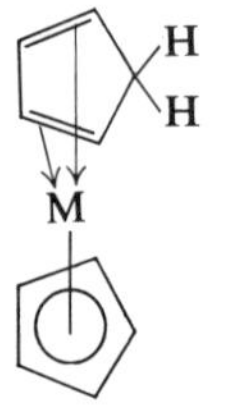

XIII 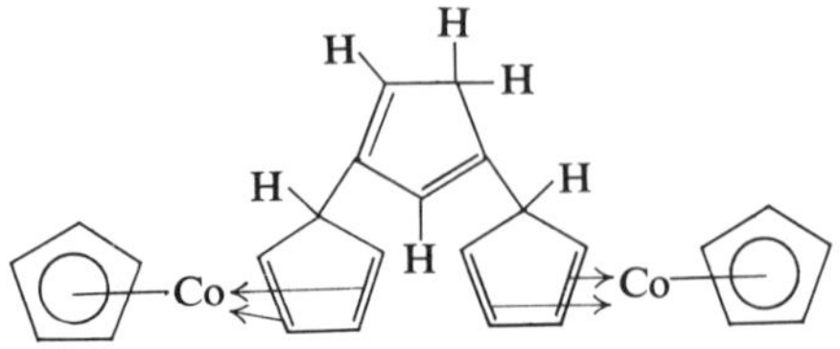   **XIV**

The chemistry of the biscyclopentadienyl derivatives of rhodium and iridium is appreciably different from that of the biscyclopentadienyl deri-vatives of cobalt. Reaction of the anhydrous trichloride of rhodium or iridium with cyclopentadienylmagnesium bromide under vigorous conditions gives the colorless biscyclopentadienylmetal(III) cations, $(C_5H_5)_2M^+$ (M = Rh or Ir), which can be conveniently isolated as their hexafluorophosphates [53]. Reaction of the anhydrous trichloride of rhodium or iridium with excess sodium cyclopentadienide gives a low yield of the corresponding π-cyclo-pentadienyl-π-cyclopentadienemetal derivative $C_5H_5MC_5H_6$ (**XIII**: M = Rh or Ir) as a yellow volatile solid [54]. Reaction of the cation $(C_5H_5)_2Rh^+$ with sodium cyclopentadienide in tetrahydrofuran solution gives a mixture of yellow π-cyclopentadienyl-π-(cyclopentadienylcyclopentadiene)-rhodium, $C_5H_5RhC_5H_5C_5H_5$ (**XV**), and pale yellow $(C_5H_5)_6Rh_2$ [55]. Thus, sodium cyclopentadienide reacts differently with $(C_5H_5)_2Rh^+$ than with $(C_5H_5)_2Co^+$. Neutral monomeric biscyclopentadienylmetal(II) derivatives of rhodium and iridium, $(C_5H_5)_2M$ (M = Rh or Ir), analogous to cobaltocene have not been isolated. Apparently, the nineteen-electron configuration is even more un-

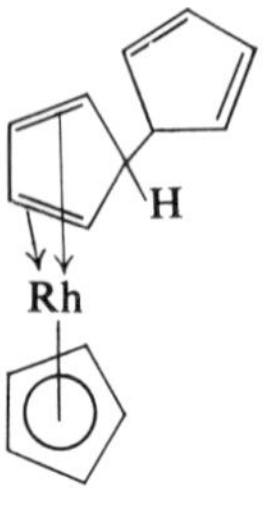

XV 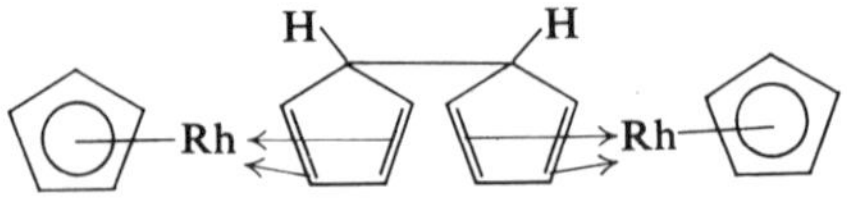   **XVI**

stable with rhodium and iridium than with cobalt. However, treatment of the salts $[(C_5H_5)_2M][PF_6]$ (M = Rh or Ir) with excess molten sodium in the absence of a solvent gives the yellow biscyclopentadienylmetal(II) dimers $(C_5H_5)_4M_2$ (M = Rh or Ir) [53]. These appear to have structure **XVI**, in which each central metal atom acquires the favored eighteen-electron rare-gas configuration by being bonded to one π-cyclopentadienyl ring and one substituted π-cyclopentadiene ring; the latter ring behaves as a conjugated diene.

Cobalt, rhodium, and iridium each form the volatile liquid cyclopentadienylmetal dicarbonyls, $C_5H_5M(CO)_2$ (**XVII**). The dark-red cobalt derivative $C_5H_5Co(CO)_2$ can be obtained either by carbonylation of $(C_5H_5)_2Co$ at elevated temperatures and, preferably, elevated pressures [56] or by reaction of $Co_2(CO)_8$ with monomeric cyclopentadiene at room temperature [57]. The orange rhodium derivative $C_5H_5Rh(CO)_2$ can be prepared by reaction of $[Rh(CO)_2Cl]_2$ with sodium cyclopentadienide in petroleum ether at room temperature [58]. The yellow iridium derivative $C_5H_5Ir(CO)_2$ can similarly be prepared by reaction of $Ir(CO)_3Cl$ with sodium cyclopentadienide at room temperature [59].

Polynuclear cyclopentadienylmetal carbonyls of cobalt and rhodium have been obtained by decarbonylation of the mononuclear $C_5H_5M(CO)_2$ (**XVII**) derivatives. Thus, ultraviolet irradiation of a hexane solution of $C_5H_5Co(CO)_2$ is reported [60] to give black trinuclear $[C_5H_5CoCO]_3$. The rhodium derivative $C_5H_5Rh(CO)_2$, upon standing, is converted to red crystalline $(C_5H_5)_2Rh_2(CO)_3$ (**XVIII**), which, upon ultraviolet irradiation, is converted to black $[C_5H_5RhCO]_3$ which can be separated into the two isomers **XIX** and **XX** [61].

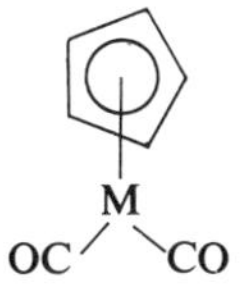

XVII 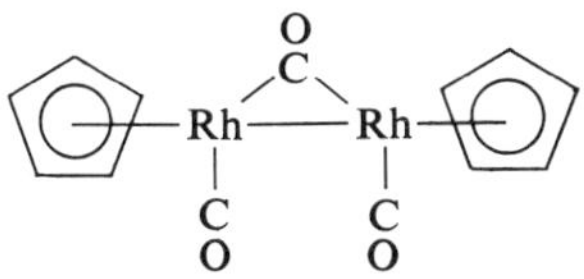   XVIII 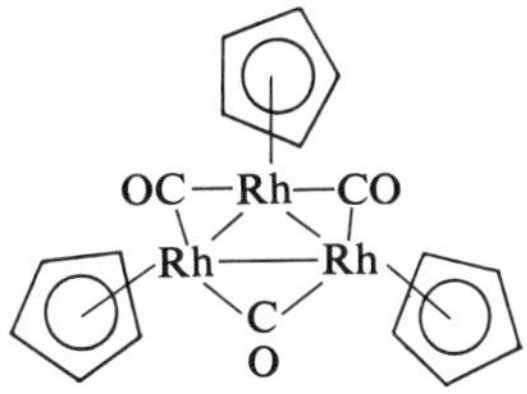   XIX

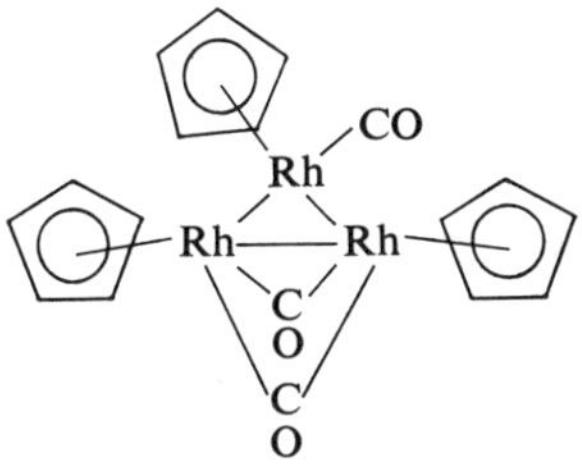

XX

A variety of reactions of $C_5H_5Co(CO)_2$ (**XVII**: M = Co) have been investigated. Many of these reactions of $C_5H_5Co(CO)_2$ resemble corresponding reactions of $Fe(CO)_5$, owing to the similarity of the C_5H_5Co and $Fe(CO)_3$ fragments. Heating $C_5H_5Co(CO)_2$ with triphenylphosphine in methylcyclohexane at 100° replaces one carbonyl group with the tertiary phosphine to give dark-red, relatively unstable $C_5H_5Co(CO)P(C_6H_5)_3$ [62]. Similarly, heating $C_5H_5Co(CO)_2$ with various diolefins at ~130° gives red to orange complexes of the type $C_5H_5Co(diene)$ (diene = butadiene, 1,3-cyclohexadiene, 1,5-cyclooctadiene, bicyclo[2,2,1]heptadiene, etc.) with displacement of both carbonyl groups [63]. The 1,3-cyclohexadiene complex $C_5H_5CoC_6H_8$ (**XXI**) is of particular interest since it undergoes a double hydride abstraction with triphenylmethyl tetrafluoroborate to give the yellow π-cyclopentadienyl-π-benzene-cobalt dication, $[C_5H_5CoC_6H_6]^{2+}$ (**XXII**), isolated as its hexafluorophosphate salt [64].

Other reactions of $C_5H_5Co(CO)_2$ are of the oxidative addition type and also parallel similar reactions of $Fe(CO)_5$. Reaction of $C_5H_5Co(CO)_2$ with iodine in diethyl ether solution gives a black precipitate of $C_5H_5Co(CO)I_2$ (**XXIII**: M = Co)[62, 65] which may be regarded as a carbonyl derivative of cobalt (III). The carbonyl group in $C_5H_5Co(CO)I_2$ (**XXIII**; M = Co) is easily replaced by Lewis bases at room temperature, giving compounds of the type $C_5H_5CoI_2L$ (L = triphenylphosphine, pyridine, etc.). Bidentate Lewis bases, such as 2,2'-bipyridyl, react with $C_5H_5Co(CO)I_2$ to give salts of the type $[C_5H_5CoIL_2]I$. Reaction of $C_5H_5Co(CO)_2$ with perfluoroalkyl iodides, R_fI, in benzene solution gives black volatile (~100°/0.1 mm) derivatives of the type $C_5H_5Co(CO)R_fI$ (**XXIV**: M = Co; $R_f = CF_3$, C_2F_5, n-C_3F_7, $(CF_3)_2CF$,

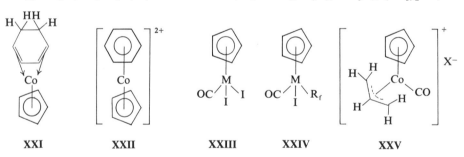

XXI	**XXII**	**XXIII**	**XXIV**	**XXV**

or n-C_7F_{15}) [63]. Reaction of $C_5H_5Co(CO)_2$ with allyl halides gives yellow salts $[C_5H_5Co(CO)(\pi$-$C_3H_5)]X$ (**XXV**) [66]; in addition, low yields of the black nonionic derivatives $C_5H_5Co(\pi$-$C_3H_5)X$ (**XXVI**) are sometimes obtained [67].

The corresponding chemistry of $C_5H_5Rh(CO)_2$ has been relatively little investigated and that of $C_5H_5Ir(CO)_2$ remains presently unknown. The π-cyclopentadienylrhodium derivatives prepared from $C_5H_5Rh(CO)_2$ include

$C_5H_5Rh(CO)I_2$ (**XXIII**: M = Rh), $C_5H_5Rh(CO)R_fI$ (**XXIV**: M = Rh, $R_f = CF_3$, C_2F_5, or $n\text{-}C_3F_7$), and the novel complex $C_5H_5RhC_6(CF_3)_6$ (**XXVII**) prepared from $C_5H_5Rh(CO)_2$ and hexafluorobutyne-2 [*68*]. This last complex **XXVII** is a unique complex of a benzene derivative where only two of the three double bonds of the benzene derivative are bonded to the metal atom.

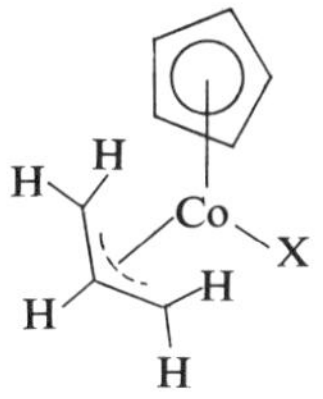

XXVI 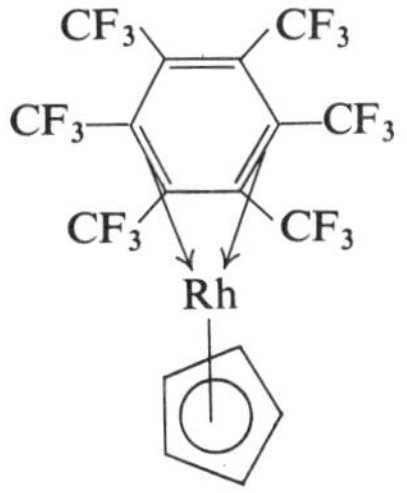   XXVII

Olefin Complexes (Rhodium)

The organometallic compounds of rhodium include a variety of olefin complexes which are of importance in rhodium-catalyzed reactions of olefins. These complexes are often easily prepared by treatment of ethanolic solutions of rhodium trichloride with the olefin or diolefin. Thus, treatment of ethanolic rhodium trichloride with excess ethylene gives the orange-red ethylene complex $[(C_2H_4)_2RhCl]_2$ (**XXVIII**) [*69*]. Nonconjugated chelating diolefins such as 1,5-cyclooctadiene, bicyclo[2,2,1]heptadiene, and 1,5-hexadiene react with ethanolic rhodium trichloride to form similar $[(\text{diene})RhCl]_2$ complexes (e.g., **XXIX** in the case of 1,5-cyclooctadiene) [*70*]. Conjugated diolefins do not form similar $[(\text{diene})RhCl]_2$ complexes with ethanolic rhodium trichloride. Instead, a few conjugated diolefins form unstable $[(\text{diene})_2RhCl]_2$ complexes where only one double bond of each conjugated diolefin is complexed to the metal atom [*71*]. In a few cases (e.g., 1,3-cyclooctadiene) a conjugated diolefin rearranges in the presence of rhodium chloride, forming a chelating nonconjugated diolefin which then forms a $[(\text{diene})RhCl]_2$ complex in the usual manner [*72*].

The chlorine bridges in compounds of the types $[(\text{olefin})_2RhCl]_2$ and $[(\text{diene})RhCl]_2$ are readily broken with suitable basic reagents. Acetylacetone in the presence of a base reacts with these rhodium complexes to form the corresponding mononuclear $(\text{olefin})_2RhC_5H_7O_2$ or $(\text{diene})RhC_5H_7O_2$ complexes. Similarly, sodium or thallium cyclopentadienide reacts with the [(olefin)$_2$RhCl]$_2$ or [(diene)RhCl]$_2$ complexes to form the corresponding C$_5$H$_5$Rh(olefin)$_2$ or C$_5$H$_5$Rh(diene) derivatives as relatively stable, yellow, volatile solids (in the cases of the ethylene and 1,5-cyclooctadiene derivatives). The ethylene derivative C$_5$H$_5$Rh(C$_2$H$_4$)$_2$ (**XXX**) exhibits a temperature-dependent NMR spectrum which arises from restricted rotation of the ethylene ligands [73].

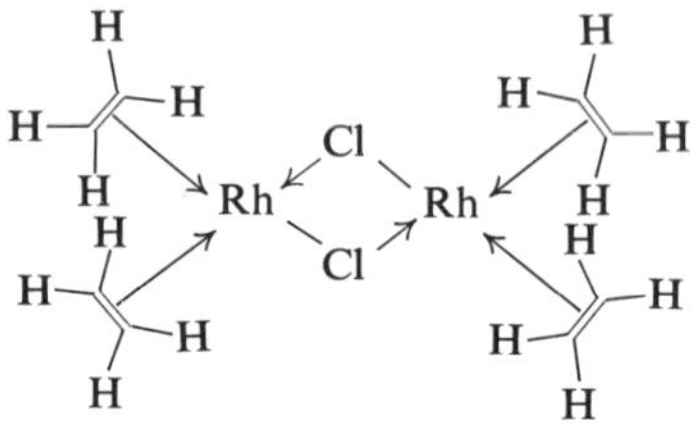

XXVIII

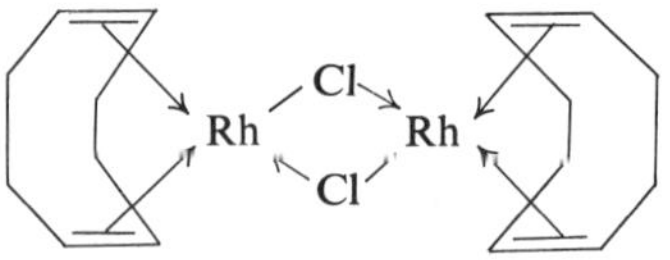

XXIX

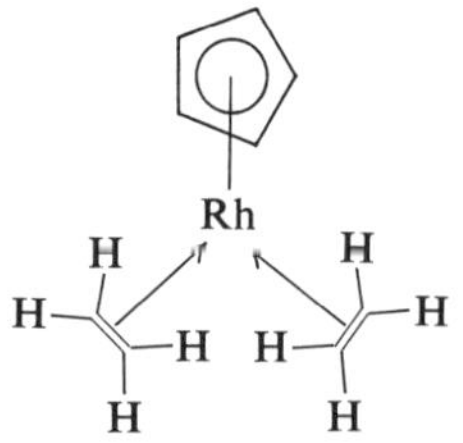

XXX

Allyl Derivatives (Rhodium)

π-Allylcobalt derivatives have already been discussed in the sections on cobalt carbonyls (p. 151). Rhodium also forms a variety of π-allyl derivatives, but none of these correspond to the cobalt derivatives previously discussed. The key intermediate for the preparation of π-allylrhodium derivatives is orange-yellow [(C$_3$H$_5$)$_2$RhCl]$_2$ (**XXXI**), which may be obtained by reaction of [Rh(CO)$_2$Cl]$_2$ with allyl chloride in the presence of water [74]. The chlorine bridges in [(C$_3$H$_5$)$_2$RhCl]$_2$ (**XXXI**) can be broken with thallium acetyl-acetonate to give the volatile mononuclear acetylacetonate (C$_3$H$_5$)$_2$RhC$_5$H$_7$O$_2$ (**XXXII**), and with thallium cyclopentadienide to give yellow volatile C$_5$H$_5$Rh(C$_3$H$_5$)$_2$ (**XXXIII**). The latter compound **XXXIII** is unique in containing both a σ-allyl and a π-allyl group. Reaction of [(C$_3$H$_5$)$_2$RhCl]$_2$ (**XXXI**) with excess allylmagnesium chloride gives yellow triallylrhodium, (C$_3$H$_5$)$_3$Rh (**XXXIV**: M = Rh,) mp 80°–85°, which exhibits a temperature-dependent proton NMR spectrum because of fluxional processes involving the three

π-allyl groups [75]. An iridium analogue $(C_3H_5)_3Ir$ (**XXXIV**: M = Ir) has been prepared in relatively low yield from allylmagnesium chloride and iridium(III) acetylacetonate [76]. Allylmagnesium chloride also cleaves the chlorine bridges in the 1,5-cyclooctadiene complex $[C_8H_{12}RhCl]_2$ to give the mononuclear, pale yellow π-allyl-π-1,5-cyclooctadiene–rhodium, $C_3H_5RhC_8H_{12}$ (**XXXV**), mp 108°–112° [77].

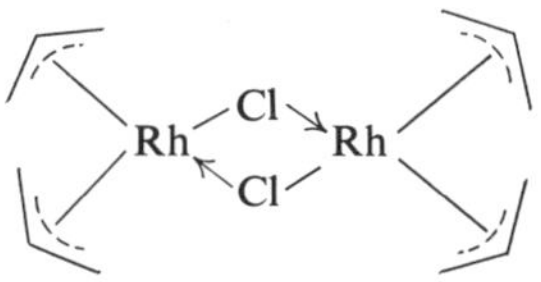

XXXI 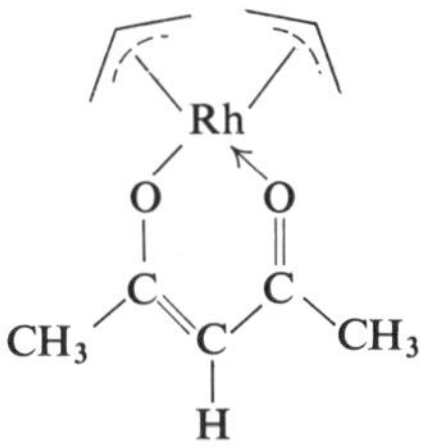 XXXI 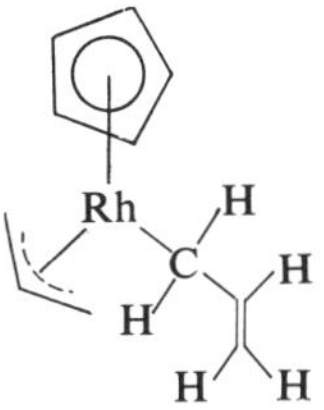 XXXIII

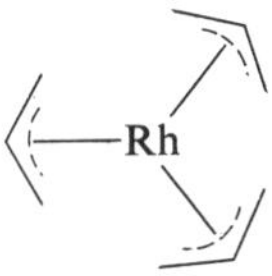

XXXIV 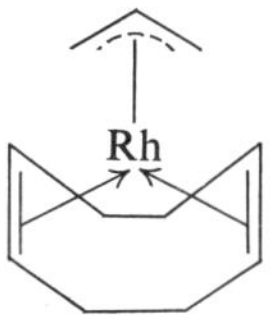 XXXV

Cobalt Alkyls

Cobalt(II) and cobalt(III) both form alkyl derivatives with cobalt–carbon σ-bonds. Additional ligands about the cobalt atom are necessary for the stability of the complexes. Most of the alkyl–cobalt derivatives are prepared from corresponding cobalt halide complexes by treatment with the reactive alkyl derivatives of lithium, magnesium, or aluminum.

The known cobalt(II) alkyls are of two types: the tertiary phosphine derivatives [78] $(R_3'P)_2CoR_2$ and the 2,2'-bipyridyl derivatives [79] $(C_{10}H_8N_2)CoR_2$. The tertiary phosphine derivatives $(R_3'P)_2CoR_2$ have only been isolated in the relatively favorable case where R is an *ortho*-substituted aryl group such as mesityl, 2-methyl-1-naphthyl or an alkynyl group such as phenylethynyl. The 2,2'-bipyridyl derivatives $(C_{10}H_8N_2)CoR_2$ are obtained from cobalt(II) acetylacetonate, 2,2'-bipyridyl, and alkylaluminum compounds.

The known cobalt(III) alkyls are more stable and extensive in scope than the known cobalt(II) alkyls. The cobalt(III) alkyls are all hexacoordinate and octahedral. They include the cyanide derivatives $[RCo(CN)_5]^{3-}$ [80], the dimethylglyoxime derivatives $[(CH_3)_2C_2N_2O_2H]_2CoLR$ (**XXXVI**: R = alkyl group, L = Lewis base such as pyridine, dimethylsulfide, or water) [81], the etioporphyrin derivatives $(CH_3)_4(C_2H_5)_4C_{20}H_4N_4CoLR$ (**XXXVII**: R = CH_3, C_2H_5, n-C_3H_7, n-C_4H_9, C_6H_5, etc.; L = pyridine) [82], and the derivatives of the Schiff base bis(acetylacetone)ethylenediimine

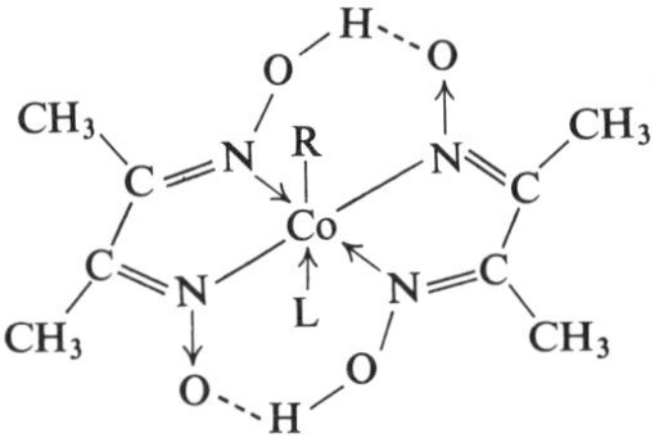

XXXVI

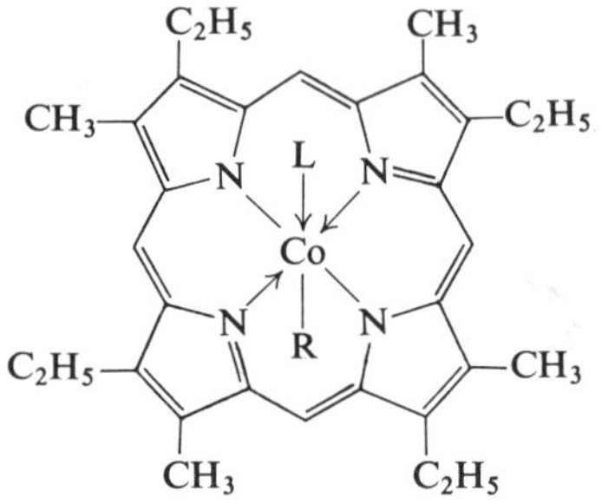

XXXVII

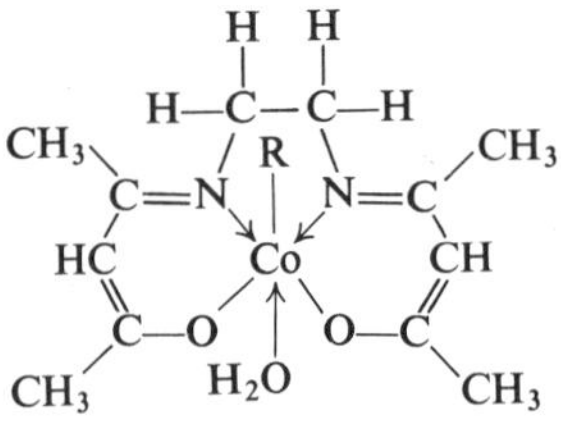

XXXVIII

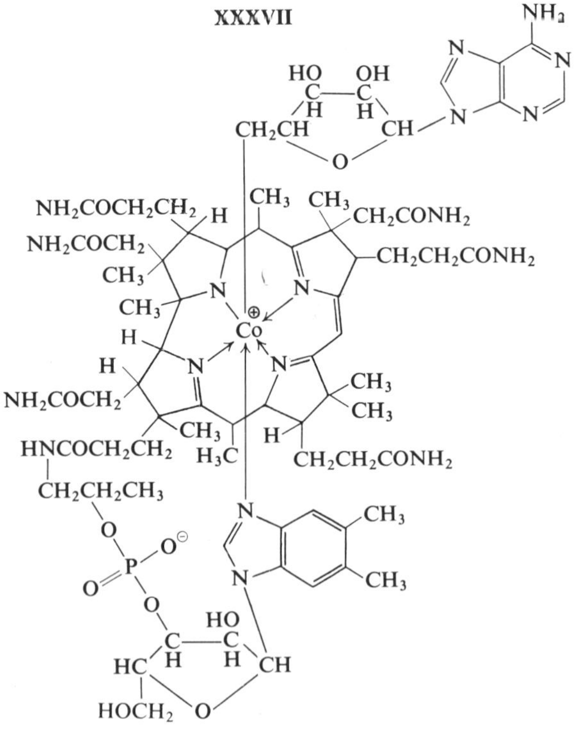

XXXIX

$(CH_3)_4C_8H_6N_2O_2Co(H_2O)R$ (**XXXVIII**: $R = CH_3$, C_2H_5, or C_6H_5) [*83*]. Most of these cobalt(III) derivatives are prepared by alkylation of the corresponding chlorides with alkylmagnesium compounds. A few are prepared by reduction of the chlorides to unstable low-valent cobalt derivatives with sodium or sodium borohydride followed by alkylation of these low-valent cobalt derivatives with alkyl halides or similar alkylating agents. The cyanide derivatives $[RCo(CN)_5]^{3-}$ are best prepared by alkylation of the cobalt(II) cyanide complex $[Co(CN)_5]^{3-}$ obtained by mixing aqueous solutions of cobalt(II) salts with excess potassium cyanide. Much of the chemistry of these cobalt(III) alkyls parallels that of the natural product vitamin B_{12} coenzyme (**XXXIX**) which may be regarded as a more complex cobalt(III) alkyl.

REFERENCES

1. I. Wender, H. W. Sternberg, S. Metlin, and M. Orchin, *Inorg. Syn.* **5**, 190 (1957); P. Szabó, L. Markó, and G. Bor, *Chem. Tech.* (*Berlin*) **13**, 549 (1961).
2. R. Ercoli, P. Chini, and M. Massi-Mauri, *Chim. Ind.* (*Milan*) **41**, 132 (1959).
3. P. Chini, *Chem. Commun.* p. 440 (1967).
4. G. G. Sumner, H. P. Klug, and L. E. Alexander, *Acta Cryst.* **17**, 732 (1964).
5. K. Noack, *Spectrochim. Acta* **19**, 1925 (1963); K. Noack, *Helv. Chim. Acta* **47**, 1064 and 1555 (1964).
6. C. H. Wei and L. F. Dahl, *J. Am. Chem. Soc.* **88**, 1821 (1966).
7. E. R. Corey, L. F. Dahl, and W. Beck, *J. Am. Chem. Soc.* **85**, 1202 (1963).
8. W. Hieber, O. Vohler, and G. Braun, *Z. Naturforsch.* **13b**, 192 (1958).
9. W. Hieber, J. Sedlmeier, and W. Abeck, *Ber.* **87**, 25 (1954).
10. W. Hieber and W. Freyer, *Ber.* **93**, 462 (1960).
11. W. Hieber and J. Sedlmeier, *Ber.* **87**, 789 (1954).
12. W. Hieber, E. O. Fischer, and E. Böckly, *Z. Anorg. Allgem. Chem.* **269**, 308 (1952).
13. H. W. Sternberg, I. Wender, R. A. Friedel, and M. Orchin, *J. Am. Chem. Soc.* **75**, 2717 (1953).
14. W. Hieber, O. Vohler, and G. Braun, *Z. Naturforsch.* **13b**, 192 (1958).
15. R. F. Heck and D. S. Breslow, *Chem. & Ind.* (*London*) p. 467 (1960).
16. W. R. McClellan, *J. Am. Chem. Soc.* **83**, 1598 (1961).
17. B. J. Aylett and J. M. Campbell, *Chem. Commun.* p. 217 (1965).
18. A. J. Chalk and J. F. Harrod, *J. Am. Chem. Soc.* **89**, 1640 (1967).
19. S. Breitschaft and F. Basolo, *J. Am. Chem. Soc.* **88**, 2702 (1966).
20. H. B. Jonassen, R. I. Stearns, J. Kentämaa, D. W. Moore, and A. G. Whittaker, *J. Am. Chem. Soc.* **80**, 2586 (1958).
21. R. F. Heck and D. S. Breslow, *J. Am. Chem. Soc.* **82**, 750 (1960).
22. R. B. King and M. B. Bisnette, *Inorg. Chem.* **3**, 785 (1964).
23. F. Seel, *Z. Anorg. Allgem. Chem.* **269**, 40 (1952).
24. R. L. Mond and A. E. Wallis, *J. Chem. Soc.* **121**, 34 (1922).
25. L. Malatesta and A. Araneo, *J. Chem. Soc.* p. 3803 (1957).
26. W. T. Dent, L. A. Duncanson, R. G. Guy, H. W. B. Reed, and B. L. Shaw, *Proc. Chem. Soc.* p. 169 (1961); G. Bor, L. Markó, and B. Markó, *Ber.* **95**, 333 (1962); R. Ercoli, S. Santambrogio, and G. Tettamanti Casagrande, *Chim. Ind.* (*Milan*) **44**, 1344 (1962).
27. R. Markby, I. Wender, R. A. Friedel, F. A. Cotton, and H. W. Sternberg, *J. Am. Chem. Soc.* **80**, 6529 (1958).

28. H. Greenfield, H. W. Sternberg, R. A. Friedel, J. H. Wotiz, R. Markby, and I. Wender, *J. Am. Chem. Soc.* **78**, 120 (1956).

29. H. W. Sternberg, J. G. Shukys, C. D. Donne, R. Markby, R. A. Friedel, and I. Wender, *J. Am. Chem. Soc.* **81**, 2339 (1959).

30. U. Krüerke, C. Hoogzand, and W. Hübel, *Ber.* **94**, 2817 (1961).

31. U. Krüerke and W. Hübel, *Ber.* **94**, 2829 (1961).

32. L. F. Dahl and D. L. Smith, *J. Am. Chem. Soc.* **84**, 2450 (1962).

33. W. Hieber and H. Lagally, *Z. Anorg. Allgem. Chem.* **251**, 96 (1943).

34. S. H. H. Chaston and F. G. A. Stone, *Chem. Commun.* p. 964 (1967).

35. C. H. Wei, G. R. Wilkes, and L. F. Dahl, *J. Am. Chem. Soc.* **89**, 4792 (1967).

36. J. A. McCleverty and G. Wilkinson, *Inorg. Syn.* **8**, 211 (1966).

37. R. F. Heck, *J. Am. Chem. Soc.* **86**, 2796 (1964).

38. L. M. Vallarino, *Inorg. Chem.* **4**, 161 (1965).

39. F. Bonati and G. Wilkinson, *J. Chem. Soc.* p. 3156 (1964).

40. P. N. Lawson and G. Wilkinson, *J. Chem. Soc.* p. 1900 (1965).

41. G. R. Wilkes, Ph.D. Thesis, University of Wisconsin, Madison, Wisconsin (1965).

42. W. Hieber and H. Lagally, *Z. Anorg. Allgem. Chem.* **245**, 321 (1940).

43. E. O. Fischer and K. S. Brenner, *Z. Naturforsch.* **17b**, 774 (1962).

44. J. P. Collman and J. W. Kang, *J. Am. Chem. Soc.* **89**, 944 (1967).

45. L. Vaska and J. W. Diluzio, *J. Am. Chem. Soc.* **84**, 679 (1962).

46. L. Vaska, *Science* **140**, 809 (1963).

47. R. S. Nyholm and K. Vrieze, *Chem. & Ind.* (*London*) p. 318 (1964).

48. G. W. Parshall and F. N. Jones, *J. Am. Chem. Soc.* **87**, 5356 (1965).

49. G. Wilkinson, F. A. Cotton, and J. M. Birmingham, *J. Inorg. & Nucl. Chem.* **2**, 95 (1956).

50. G. Wilkinson, *J. Am. Chem. Soc.* **74**, 6148 (1952).

51. M. L. H. Green, L. Pratt, and G. Wilkinson, *J. Chem. Soc.* p. 3753 (1959).

52. E. O. Fischer, W. Fellmann, and G. E. Herberich, *Ber.* **95**, 2254 (1962).

53. E. O. Fischer and H. Wawersik, *J. Organometal. Chem.* (*Amsterdam*) **5**, 559 (1966).

54. M. L. H. Green, L. Pratt, and G. Wilkinson, *J. Chem. Soc.* p. 3753 (1959).

55. R. J. Angelici and E. O. Fischer, *J. Am. Chem. Soc.* **85**, 3733 (1963).

56. E. O. Fischer and R. Jira, *Z. Naturforsch.* **10b**, 355 (1955).

57. T. S. Piper, F. A. Cotton, and G. Wilkinson, *J. Inorg. & Nucl. Chem.* **1**, 165 (1955).

58. E. O. Fischer and K. Bittler, *Z. Naturforsch.* **16b**, 225 (1961).

59. E. O. Fischer and K. S. Brenner, *Z. Naturforsch.* **17b**, 774 (1962).

60. R. B. King, *Inorg. Chem.* **5**, 2227 (1966).

61. O. S. Mills and E. F. Paulus, *Chem. Commun.* p. 815 (1966); O. S. Mills and E. F. Paulus, *J. Organometal. Chem.* **10**, 331 (1967); E. F. Paulus, E. O. Fischer, H. P. Fritz, and H. Schuster-Woldan, *ibid.* p. P3.

62. R. B. King, *Inorg. Chem.* **5**, 82 (1966).

63. R. B. King, P. M. Treichel, and F. G. A. Stone, *J. Am. Chem. Soc.* **83**, 3600 (1961).

64. E. O. Fischer and R. D. Fischer, *Z. Naturforsch.* **16b**, 556 (1961).

65. R. F. Heck, *Inorg. Chem.* **4**, 855 (1965).

66. E. O. Fischer and R. D. Fischer, *Z. Naturforsch.* **16b**, 475 (1961).

67. R. F. Heck, *J. Org. Chem.* **28**, 604 (1963).

68. R. S. Dickson and G. Wilkinson, *J. Chem. Soc.* p. 2699 (1964); J. A. McCleverty and G. Wilkinson, *ibid.* p. 4200.

69. R. D. Cramer, *Inorg. Chem.* **1**, 722 (1962).

70. J. Chatt and L. M. Venanzi, *J. Chem. Soc.* p. 4735 (1957); E. W. Abel, M. A. Bennett, and G. Wilkinson, *ibid.* p. 3178 (1959).

71. L. Porri, A. Lionetti, G. Allegra, and A. Immirzi, *Chem. Commun.* p. 336 (1965); L. Porri and A. Lionetti, *J. Organometal. Chem. (Amsterdam)* **6**, 422 (1966).
72. R. E. Rinehart and J. S. Lasky, *J. Am. Chem. Soc.* **86**, 2516 (1964).
73. R. D. Cramer, *J. Am. Chem. Soc.* **86**, 217 (1964).
74. J. Powell and B. L. Shaw, *J. Chem. Soc. A* p. 583 (1968).
75. J. K. Becconsall and S. O'Brien, *Chem. Commun.* p. 720 (1966).
76. P. Chini and S. Martinengo, *Inorg. Chem.* **6**, 837 (1967).
77. A. Kasahara and K. Tanaka, *Bull. Chem. Soc. Japan* **39**, 634 (1966).
78. J. Chatt and B. L. Shaw, *J. Chem. Soc.* p. 1718 (1960); p. 285 (1961).
79. T. Saito, Y. Uchida, A. Misono, A. Yamamoto, K. Morituji, and S. Ikeda, *J. Organometal. Chem. (Amsterdam)* **5**, 493 (1966).
80. J. Halpern and J. P. Maher, *J. Am. Chem. Soc.* **86**, 2311 (1964); J. Kwiatek and J. K. Seyler, *J. Organometal. Chem.* **3**, 421 and 433 (1965).
81. G. N. Schrauzer and R. J. Windgassen, *J. Am. Chem. Soc.* **88**, 3738 (1966).
82. D. Dolphin and A. W. Johnson, *Chem. Commun.* p. 494 (1965); D. A. Clarke, R. Grigg, and A. W. Johnson, *ibid.* p. 208 (1966).
83. G. Costa, G. Mestroni, G. Tauzher, and L. Stefani, *J. Organometal. Chem. (Amsterdam)* **6**, 181 (1966).

SUPPLEMENTARY READING

1. R. F. Heck, Synthesis and reactions of alkylcobalt and acylcobalt tetracarbonyls. *Advan. Organometal. Chem.* **4**, 243 (1966).
2. A. J. Chalk and J. F. Harrod, Catalysis by cobalt carbonyls. *Advan. Organometal. Chem.* **6**, 119 (1968).
3. G. N. Schrauzer, Organocobalt chemistry of vitamin B_{12} model compounds (cobaloximes). *Accounts Chem. Res.* **1**, 97 (1968).
4. R. D. Cramer, Transition metal catalysis exemplified by some rhodium-promoted reactions of olefins. *Accounts Chem. Res.* **1**, 186 (1968).
5. L. Vaska, Reversible activation of covalent molecules by transition metal complexes. The role of the covalent molecule. *Accounts Chem. Res.* **1**, 335 (1968).

QUESTIONS

1. Suggest a reason for the presence of bridging carbonyls in $Co_2(CO)_8$, but the absence of bridging carbonyls in $Mn_2(CO)_{10}$.
2. Suggest reasons for the greater stability of silylcobalt tetracarbonyls, $R_3SiCo(CO)_4$, than for alkylcobalt tetracarbonyls such as $CH_3Co(CO)_4$.
3. Show how the chemistry of the biscyclopentadienyl derivatives of the three $5d$ transition metals, rhenium, osmium, and iridium, differs greatly and how these differences can be related to the tendency for transition metals to acquire the favored eighteen-electron rare-gas configuration in their π-cyclopentadienyl derivatives.

CHAPTER VII

Organometallic Derivatives of Nickel, Palladium, and Platinum

Introduction

Neutral nickel, palladium, and platinum each require eight electrons from the surrounding ligands to attain the normally favored eighteen-electron rare-gas configuration. Tetracarbonylnickel, $Ni(CO)_4$, and its substitution products thus have the rare-gas configuration. In addition π-cyclopentadienylmetal derivatives of the type C_5H_5MT (M = Ni, Pd, or Pt; T = three-electron donor ligand) also have the eighteen-electron rare-gas configuration.

In the case of most transition metals, the vast majority of organometallic derivatives have the favored eighteen-electron rare-gas configuration. However, most organopalladium and organoplatinum derivatives do not have this rare-gas electronic configuration. Instead, square planar organopalladium(II) and organoplatinum(II) derivatives with only sixteen-electron configurations are frequently found. Another exception to the "rare-gas rule" is biscyclopentadienylnickel (nickelocene) $(C_5H_5)_2Ni$, a stable green solid which has a twenty-electron configuration, which is two in excess of the rare-gas electronic configuration.

Organometallic Chemistry of Nickel

Nickel forms a tetracarbonyl, $Ni(CO)_4$, which is a colorless, extremely volatile (bp $43°/760$ mm) liquid which decomposes to metallic nickel and carbon monoxide upon heating above room temperature. The toxicity of $Ni(CO)_4$ is comparable to that of HCN, making hazardous the handling of $Ni(CO)_4$. Tetracarbonylnickel can be prepared by reaction of finely divided nickel with carbon monoxide at atmospheric pressure [1]. This facile preparation of $Ni(CO)_4$ was a major reason why this carbonyl was the first metal carbonyl derivative to be prepared [2]. Furthermore, the conversion of metallic nickel to volatile $Ni(CO)_4$ by treatment with carbon monoxide at atmospheric pressure is the basis for the Mond process for separating nickel from other metals; pyrolysis of $Ni(CO)_4$ regenerates the nickel metal in a pure form. An alternate method for preparing $Ni(CO)_4$ utilizes the reduction of nickel salts with dithionite ion in aqueous ammonia solution in the presence of carbon monoxide at atmospheric pressure [3].

Tricovalent phosphorus derivatives and related ligands react with $Ni(CO)_4$ to form substitution products of the general type $L_nNi(CO)_{4-n}$. As carbonyl groups in $Ni(CO)_4$ become replaced with more weakly π-accepting tertiary phosphine or isocyanide ligands, the remaining carbonyl groups become progressively more difficult to replace with such ligands. However, in the cases of isocyanides and tertiary phosphines, complete substitution of the four carbonyl groups in $Ni(CO)_4$ has been achieved to give carbonyl-free products of the type NiL_4.

Another important organonickel derivative is biscyclopentadienylnickel (nickelocene), a dark-green volatile (sublimes at $50°/0.1$ mm) crystalline solid [4]. Nickelocene has a twenty-electron configuration, which is two greater than that of the next rare gas. These two extra electrons are unpaired, as indicated by the magnetic moment of 2.86 BM. Oxidation of nickelocene in acidic solution removes one of these "extra" electrons, giving the unstable deep-yellow cation $(C_5H_5)_2Ni^+$ with a nineteen-electron configuration. However, attempts to remove the second "extra" electron lead to complete decomposition rather than formation of a dication $(C_5H_5)_2Ni^{2+}$ which would have the favored eighteen-electron rare-gas configuration. Nickelocene can be prepared either by reaction of hexamminenickel(II) chloride with sodium cyclopentadienide in tetrahydrofuran solution [method (a) [5]] or by treatment of a mixture of nickel(II) bromide and diethylamine with cyclopentadiene [method (b) [6]]:

$$Ni(NH_3)_6Cl_2 + 2\,NaC_5H_5 \xrightarrow{\text{THF}} (C_5H_5)_2Ni + 2\,NaCl + 6\,NH_3 \qquad \text{(a)}$$

$$NiBr_2 + 2\,Et_2NH + 2\,C_5H_6 \longrightarrow (C_5H_5)_2Ni + 2\,[Et_2NH_2]Br \qquad \text{(b)}$$

The nickel(II) bromide required for method (b) is conveniently prepared by bromination of nickel metal powder in 1,2-dimethoxyethane solution and used without isolation.

Nickelocene is rather reactive, with a tendency to be converted from a derivative with a twenty-electron configuration to various derivatives with the more favored eighteen-electron rare-gas configuration. Thus, nickelocene can be reduced with sodium amalgam in methanol to give red, very volatile π-cyclopentadienyl-π-cyclopentenyl nickel, $C_5H_5NiC_5H_7$ (I), containing a three-electron π-allylic C_5H_7 ligand and a five-electron π-cyclopentadienyl

C_5H_5 ligand [7]. Treatment of nickelocene with nitric oxide at room temperature and atmospheric pressure gives dark-red liquid π-cyclopentadienyl-nitrosylnickel, C_5H_5NiNO (II), bp 144°–145°/715 mm [8]. This is a rare example of a π-cyclopentadienylmetal derivative C_5H_5ML with only one other monodentate ligand. Reaction of nickelocene with thiols proceeds according to the following equation to give the brown-black binuclear derivative $[C_5H_5NiSR]_2$ (III) [9]:

$$2\,(C_5H_5)_2Ni + 2\,RSH \longrightarrow [C_5H_5NiSR]_2 + 2\,C_5H_6$$

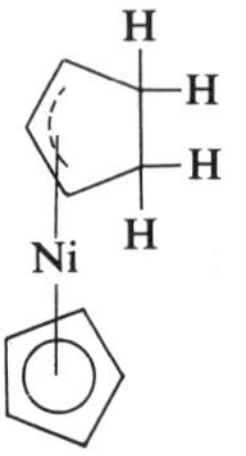

I 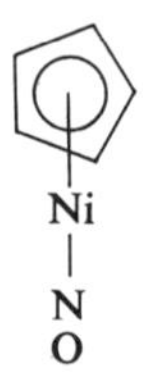   II 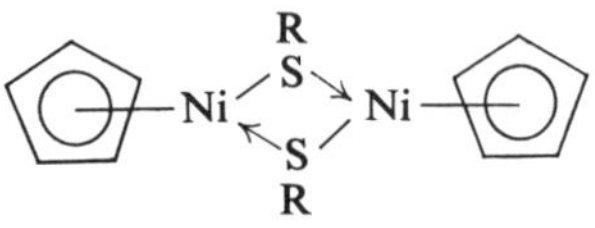   III

Nickelocene also reacts with other nickel complexes to form π-cyclopentadienylnickel derivatives, also containing the ligands in the other nickel complexes. Thus, heating nickelocene with $Ni(CO)_4$ in benzene at 80° gives red-violet crystalline $[C_5H_5NiCO]_2$ (IV) according to the following equation [10]:

$$(C_5H_5)_2Ni + Ni(CO)_4 \xrightarrow[\text{80°}]{\text{benzene}} [C_5H_5NiCO]_2 + 2\,CO$$

At higher temperatures nickelocene and $Ni(CO)_4$ form the green-brown trinuclear $(C_5H_5)_3Ni_3(CO)_2$ with structure V containing two three-way bridging carbonyl groups. Nickelocene reacts with the tertiary phosphine nickel halides $(R_3P)_2NiX_2$ to give the dark-red complexes $C_5H_5Ni(PR_3)X$ (VI: X = Cl, Br, or I) according to the following equation [11]:

$$(C_5H_5)_2Ni + (R_3P)_2NiX_2 \longrightarrow 2\,C_5H_5Ni(PR_3)X$$

These $C_5H_5Ni(PR_3)X$ derivatives react with organomagnesium compounds or other reactive organometallic derivatives to give stable green compounds of the type $C_5H_5Ni(PR_3)R'$ (VI: $R' = CH_3$, C_2H_5, C_6H_5, etc.) containing nickel–carbon σ-bonds.

A further characteristic of nickelocene is the ease of replacement of its two π-cyclopentadienyl ligands with ligands capable of forming stable nickel(0) complexes. This permits the preparation of completely substituted nickel(0) complexes of the type NiL_4 without using the highly toxic $Ni(CO)_4$. Ligands which react with nickelocene to form NiL_4 complexes include phosphorus trifluoride, trialkylphosphites, and carbon monoxide [12].

IV

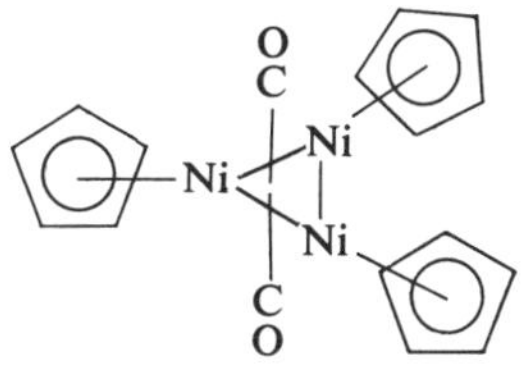

V

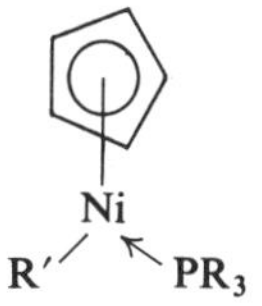

VI

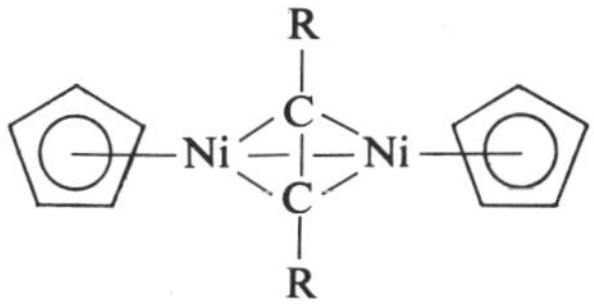

VII

Reactions of $(C_5H_5)_2Ni$ or $[C_5H_5NiCO]_2$ with various alkynes yield two types of products of interest [13]. Most alkynes form green complexes of the type (alkyne) $[NiC_5H_5]_2$ of structure VII analogous to that of the (alkyne)$Co_2(CO)_6$ compounds (p. 152) but with C_5H_5Ni groups replacing the approximately "isoelectronic" $Co(CO)_3$ groups. In addition, alkynes with electronegative substituents such as $CF_3C{\equiv}CCF_3$ and $CH_3CO_2C{\equiv}CCO_2CH_3$ undergo an addition reaction with $(C_5H_5)_2Ni$ to give orange complexes of the type $C_5H_5NiC_7H_5R_2$ (VIII: $R = CF_3$ or CO_2CH_3) containing a chelating bicyclo[2,2,1]heptadienyl ligand.

A few other π-cyclopentadienylnickel compounds of interest can be obtained from $[C_5H_5NiCO]_2$ (IV). Reaction of $[C_5H_5NiCO]_2$ with iodine at low temperatures gives black $C_5H_5Ni(CO)I$ which decomposes at 0°. A similar reaction of $[C_5H_5NiCO]_2$ with perfluoroalkyl iodides (R_fI) at room temperature gives $C_5H_5Ni(CO)I$ (which decomposes) and the dark-red volatile liquid $R_fNi(CO)C_5H_5$ derivatives (IX: $R_f = CF_3, C_2F_5$, and C_3F_7) [14]. The analogous, but extremely unstable, red liquid, $CH_3Ni(CO)C_5H_5$, can be obtained in low yield by reduction of $[C_5H_5NiCO]_2$ with potassium amalgam in tetrahydrofuran to give a solution containing $K[C_5H_5NiCO]$, followed by treatment of this solution with methyl iodide. The much higher stability of the $R_fNi(CO)C_5H_5$ derivatives than of $CH_3Ni(CO)C_5H_5$ is a further indication of the stability of perfluoroalkyl–transition metal derivatives as compared with that of nonfluorinated alkyl–transition metal derivatives. Reaction of $[C_5H_5NiCO]_2$ with stannous chloride in boiling tetrahydrofuran results in insertion of the tin atom into the nickel–nickel bond, giving green $C_5H_5Ni(CO)SnCl_2Ni(CO)C_5H_5$ (X) [15]. These $C_5H_5Ni(CO)Y$ compounds prepared from $[C_5H_5NiCO]_2$ react with triphenylphosphine at room temperature to form the more stable and crystalline green $C_5H_5Ni[P(C_6H_5)_2]Y$ compounds (VI: $R = C_6H_5$).

Nickel forms several interesting olefin complexes. Reaction of a mixture of 1,5,9-cyclododecatriene and nickel(II) acetylacetonate with diethylethoxyaluminum in diethyl ether solution gives red cyclododecatriene–nickel, $C_{12}H_{18}Ni$, as a very air-sensitive solid subliming at $40°/0.0001$ mm [16]. X-ray crystallography indicates $C_{12}H_{18}Ni$ to have structure XI with the nickel atom located in the center of the twelve-membered ring [17]. The nickel atom in $C_{12}H_{18}Ni$ (XI) has only a sixteen-electron configuration, which suggests that this complex might be quite reactive. This is indeed the case. Thus $C_{12}H_{18}Ni$ reacts with 1,5-cyclooctadiene at room temperature with replacement of the 1,5-cyclododecatriene ligand to give the yellow 1,5-cyclooctadiene complex, $(C_8H_{12})_2Ni$ (XII: $M = Ni$), which may also be obtained by treatment of a mixture of 1,5-cyclooctadiene and nickel(II) acetylacetonate with triethylaluminum in benzene solution in the presence of catalytic quantities of butadiene [16]. The complex $(C_8H_{12})_2Ni$ (XII: $M = Ni$) has the favored eighteen-electron configuration and is more stable than the complex $C_{12}H_{18}Ni$ (XI) which has a sixteen-electron configuration. Reaction of $C_{12}H_{18}Ni$ (XI) with cyclooctatetraene leads to a brown insoluble polymer $[C_8H_8Ni]_n$ which appears to have structure XIII.

VIII   IX   X   XI

The olefin complexes $C_{12}H_{18}Ni$ (XI), $(C_8H_{12})_2Ni$ (XII: $M = Ni$), and $[C_8H_8Ni]_n$ (XIII) do not appear to be preparable by displacement of the carbonyl groups from $Ni(CO)_4$ with the appropriate olefin. However, duroquinone, which may be regarded as a diolefin, reacts with $Ni(CO)_4$ to displace all four carbonyl groups giving red air-stable bisduroquinone–nickel, $[(CH_3)_4C_6O_2]_2Ni$ (XIV) [17]. Reaction of bisduroquinone–nickel (XIV) with chelating nonconjugated diolefins (e.g., 1,5-cyclooctadiene, bicyclo[2,2,1]-heptadiene, dicyclopentadiene, or two double bonds of cyclooctatetraene) replaces one of the two duroquinone ligands, giving red complexes of the type $(CH_3)_4C_6O_2Ni(diene)$ [18].

Tetramethylcyclobutadiene also forms a nickel complex. Reaction of 1,2,3,4-tetramethyl-3,4-dichlorocyclobutene, $(CH_3)_4C_4Cl_2$, with $Ni(CO)_4$ in benzene solution gives purple air-stable tetramethylcyclobutadienedichloronickel dimer, $[(CH_3)_4C_4NiCl_2]_2$ (**XV**), according to the following equation [19]:

$$2\ (CH_3)_4C_4Cl_2 + 2\ Ni(CO)_4 \longrightarrow [(CH_3)_4C_4NiCl_2]_2 + 8\ CO$$

Upon pyrolysis at $\sim 200°$ this complex (**XV**) gives tetramethylcyclobutadiene dimers ($C_{16}H_{24}$ isomers) and nickel (II) chloride. Reaction of $[(CH_3)_4C_4NiCl_2]_2$ (**XV**) with sodium cyclopentadienide gives a red complex of composition $C_5H_5NiC_{13}H_{17}$, shown to be the π-cyclobutenyl derivative **XVI** [20].

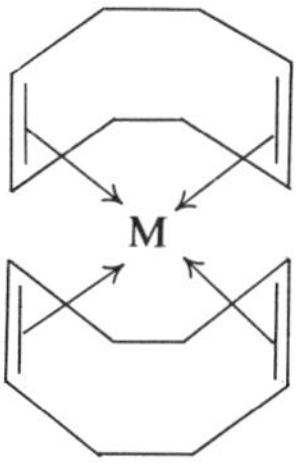

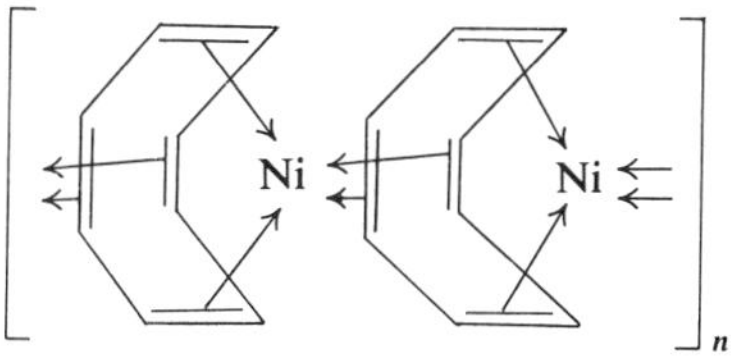

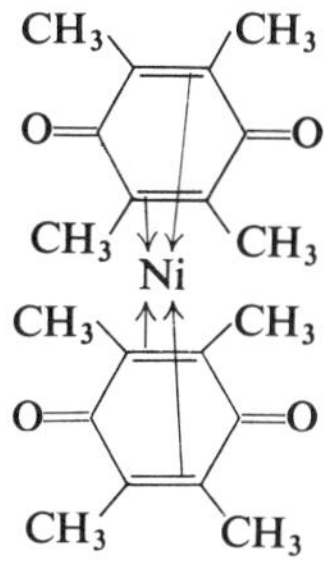

Other organonickel compounds of interest can be obtained from $Ni(CO)_4$. Thus, reaction of hexafluorobutyne-2 with $Ni(CO)_4$ at $50°$ gives the purple, crystalline, volatile $[(CF_3)_2C_2]_3Ni_4(CO)_3$, which appears to contain a novel tetrahedral cluster of nickel atoms [21]. Another derivative shown by x-ray crystallography [22] to have a tetrahedral cluster of nickel atoms is the orange tris(2-cyanoethyl)phosphine derivative $[(NCCH_2CH_2)_3P]_4Ni_4(CO)_6$ prepared from $Ni(CO)_4$ and tris(2-cyanoethyl)phosphine. Acrylonitrile reacts with $Ni(CO)_4$ to give red, nonvolatile, air-sensitive $[(CH_2=CHCN)_2Ni]_n$ of unknown molecular complexity [23].

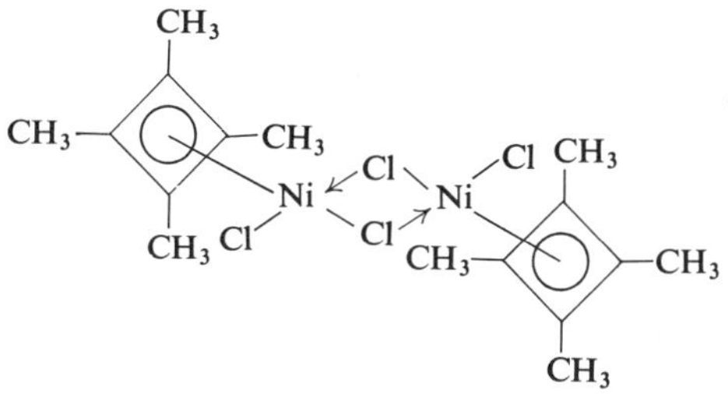

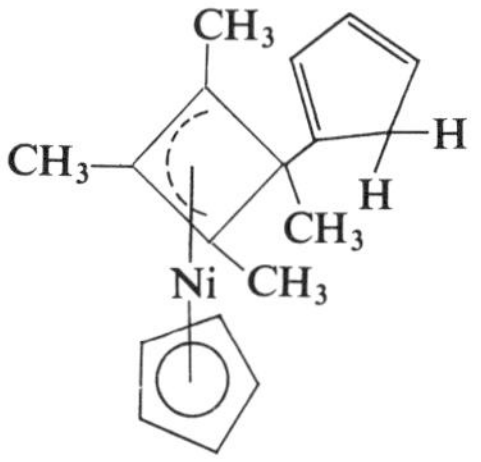

π-Allyl Derivatives of Nickel, Palladium, and Platinum

A variety of π-allyl derivatives of these metals can be prepared. Thus, reaction of the anhydrous metal dihalides with an ethereal solution of allyl-magnesium chloride gives the volatile, extremely air-sensitive bis(π-allyl)metal derivatives $(C_3H_5)_2M$ (M = Ni, yellow; M = Pd, pale yellow; M = Pt, colorless) [24]. A temperature dependence study of the NMR spectrum of the nickel derivative $(C_3H_5)_2Ni$ indicates the presence of both the *cis*-isomer **XVIIa** and the *trans*-isomer **XVIIb** [25]. Reactions of $Ni(CO)_4$ with allyl halides give the dark red, somewhat air-sensitive allylnickel halide dimers, $[C_3H_5NiX]_2$ (**XVIII**: M = Ni; X = Cl, Br, or I), according to the following equation [26]:

$$2 Ni(CO)_4 + 2 C_3H_5X \longrightarrow [C_3H_5NiX]_2 + 8 CO$$

The palladium analogues $[C_3H_5PdX]_2$ are more stable and readily prepared. Thus, yellow $[C_3H_5PdCl]_2$ (**XVIII**: M = Pd; X = Cl) can be readily prepared by boiling palladium(II) chloride in allyl alcohol [27].

Substituted π-allylic derivatives of nickel and palladium can also be prepared. Often, these may be obtained by using substituted allylic halides in syntheses analogous to those described in the last paragraph. Furthermore, $[(\pi\text{-allylic})PdX]_2$ compounds are so easily formed that they can often be prepared by heating palladium halides with unsaturated hydrocarbons (olefins, allenes, conjugated dienes, etc.) which can form π-allylic ligands by loss or gain of a hydrogen atom. For this reason, the range of π-allylic derivatives of palladium is much greater than that of nickel or platinum. Indeed, platinum halides appear to always form olefin complexes rather than π-allylic derivatives when treated with unsaturated hydrocarbons or allyl derivatives (except allylmagnesium halides).

Some organometallic derivatives of nickel, palladium, and platinum containing both π-allyl and π-cyclopentadienyl ligands have been prepared. Reaction of the halides $[C_3H_5MX]_2$ (**XVIII**: M = Ni or Pd) with sodium (or thallium) cyclopentadienide gives the corresponding $C_3H_5MC_5H_5$ derivatives (**XIX**: M = Ni or Pd). The nickel compound $C_3H_5NiC_5H_5$ is an

XVIIa   XVIIb   XVIII   XIX air-sensitive, very volatile, dark-red liquid [28]. The palladium analogue, $C_3H_5PdC_5H_5$, is a much less air-sensitive (but also less thermally stable), very volatile, dark-red solid [29]. Reaction of the propene–platinum complex $[C_3H_6PtCl_2]_2$ with a mixture of allylmagnesium chloride and cyclopenta-dienylmagnesium chloride is reported to give the yellow platinum analogue $C_3H_5PtC_5H_5$ (**XIX**: M = Pt) [30].

Alkyl Derivatives of Nickel, Palladium, and Platinum

A variety of hexacoordinate trimethylplatinum(IV) derivatives have been prepared. Reaction of K_2PtCl_6 or $PtCl_4$ with methylmagnesium iodide gives yellow trimethylplatinum iodide tetramer, $[(CH_3)_3PtI]_4$, mp 215°, shown by x-ray crystallography to have the cubic structure **XX** in which each of the platinum(IV) atoms attains the necessary coordination number of six by bonding to three iodine atoms [31].

Reactions of $[(CH_3)_3PtI]_4$ (**XX**) can involve replacement of the iodine atoms, breaking the iodine bridges, or both. The iodide ligand in $[(CH_3)_3PtI]_4$ (**XX**) can be replaced by other anionic ligands (e.g., sulfate, nitrate, or hydroxide) by treatment with the appropriate silver salt in a coordinating solvent, generally water. Many of these trimethylplatinum derivatives (e.g., the hydroxide $[(CH_3)_3PtOH]_4$) are tetrameric, with structures similar to that of the iodide **XX**. Reaction of $[(CH_3)_3PtI]_4$ with bidentate chelating diamines (e.g., 2,2′-bipyridyl or ethylenediamine) breaks up the iodine bridges of the tetrameric system giving monomeric derivatives of the type $(CH_3)_3PtI(di-amine)$ Reaction of $[(CH_3)_3PtI]_4$ with sodium cyclopentadienide both removes the iodine atoms and breaks up the tetrameric system, giving white, very volatile, air-stable cyclopentadienyltrimethylplatinum, $(CH_3)_3PtC_5H_5$, apparently with structure **XXI** [32].

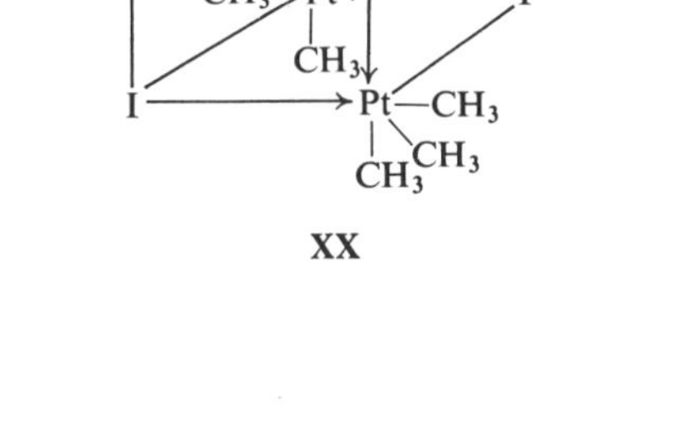

XX    XXI

Alkyl derivatives of nickel and palladium are much less stable than those of platinum. Methyl derivatives of nickel and palladium analogous to those of platinum just discussed have not been prepared and appear too unstable for preparation, probably because of the 4+ oxidation state of the metal atom. However, alkyl derivatives of all three metals can be prepared in which the metal is in the more favored 2+ oxidation state [33]. Reaction of the square planar metal(II) complexes L_2MX_2 with appropriate alkylmagnesium halides or alkyllithium compounds in excess gives the yellow to white σ-alkyl derivatives of the type L_2MR_2 (M = Ni; L = tertiary phosphine; R = ortho-substituted phenyl or substituted ethynyl groups; M = Pd; L = tertiary phosphine, tertiary arsine, dialkyl sulfide, or 1,5-cyclooctadiene; R = methyl, phenyl, phenylethynyl, or pentafluorophenyl; M = Pt; L = tertiary phosphine; R = most alkyl or aryl groups). Use of less alkylating reagent or milder reaction conditions converts the L_2MX_2 compounds into the mixed alkyl–halide derivatives L_2MXR. The stability of σ-alkyl derivatives of the types L_2MR_2 and L_2MXR increases in the following sequence: M = Ni (least stable) < Pd < Pt (most stable).

Olefin Complexes of Palladium and Platinum

The first organometallic compound to be prepared was Zeise's salt $K[C_2H_4PtCl_3]$, obtained by treatment of chloroplatinic acid with ethanol in the presence of potassium chloride [34]. A better preparation of $K[C_2H_4PtCl_3]$ utilizes the reaction of K_2PtCl_4 with ethylene in the presence of hydrochloric acid; a by-product of this reaction is $[C_2H_4PtCl_2]_2$ of structure **XXII** with chlorine bridges [35]. Reaction of $[C_2H_4PtCl_2]_2$ with excess ethylene at $-70°$ in acetone solution breaks the chlorine bridges to give yellow mono-nuclear $(C_2H_4)_2PtCl_2$ (**XXIII**) with two ethylenes bonded to the platinum atom [36]. On warming **XXIII** above $-6°$, half of its ethylene is lost and it reverts to the binuclear $[C_2H_4PtCl_2]_2$ (**XXII**).

XXII   **XXIII**

The preparation of olefin derivatives of palladium is somewhat more difficult than the preparation of analogous olefin derivatives of platinum, owing to the lower stability of the palladium–olefin bond than the platinum–olefin bond, and to the greater tendency for palladium than for platinum to form π-allylic complexes. However, palladium complexes of the type [(olefin)$PdCl_2$]$_2$ (olefin = cyclohexene, styrene, ethylene, isobutene, etc.) can be prepared by reaction of bis(benzonitrile)dichloropalladium(II), $(C_6H_5CN)_2PdCl_2$, with the olefin in an inert solvent according to the following equation [37]:

$$2 (C_6H_5CN)_2PdCl_2 + 2 \text{ olefin} \longrightarrow [(\text{olefin})PdCl_2]_2 + 4 C_6H_5CN$$

The olefin–palladium dichlorides are less stable than their platinum analogues. Unlike their platinum analogues, the olefin–palladium dichlorides are decomposed by water to give aldehydes and palladium metal. Thus, the ethylene complex [$C_2H_4PdCl_2$]$_2$ reacts with water to give acetaldehyde according to the following equation [38]:

$$[C_2H_4PdCl_2]_2 + 2 H_2O \longrightarrow 2 CH_3CHO + Pd + 4 HCl$$

Reactions of this type not only account for the inability to prepare olefin derivatives of palladium in aqueous media, but also provide the basis for the commercially important palladium-catalyzed oxidation of olefins to aldehydes [38].

Diolefins also form complexes with palladium and platinum. Reaction of chelating diolefins such as 1,5-cyclooctadiene, norbornadiene, or dicyclopentadiene with K_2PtCl_4 in alcoholic solution gives the corresponding (diene)$PtCl_2$ complexes [39]. The palladium analogues to these complexes can be obtained by reaction of the diolefin with $(C_6H_5CN)_2PdCl_2$ in an inert solvent [40].

All of these olefin and diolefin derivatives of palladium and platinum have the metal in the 2+ oxidation state. It is also possible to make a zerovalent platinum diolefin complex, bis(1,5-cyclooctadiene)platinum, $(C_8H_{12})_2Pt$ (**XII**: M = Pt), by the following sequence of reactions [41]:

$$C_8H_{12}PtCl_2 + 2 (CH_3)_2CHMgBr \longrightarrow C_8H_{12}Pt[CH(CH_3)_2]_2 + MgBr_2 + MgCl_2 \quad \text{(a)}$$

$$C_8H_{12}Pt[CH(CH_3)_2]_2 + C_8H_{12} \xrightarrow{\text{UV}} (C_8H_{12})_2Pt + 2 \{(CH_3)_2CH\cdot\} \quad \text{(b)}$$

Related zerovalent platinum complexes of unsaturated molecules may be prepared from tetrakis(triphenylphosphine)platinum(0), [$(C_6H_5)_3P]_4Pt$, by reactions such as the following:

$$[(C_6H_5)_3P]_4Pt + RC\equiv CR' \longrightarrow [(C_6H_5)_3P]_2Pt(RC_2R') + 2 (C_6H_5)_3P \quad [42] \quad \text{(a)}$$
$$\text{(XXIV)}$$

$$[(C_6H_5)_3P]_4Pt + CF_2{=}CFX \longrightarrow [(C_6H_5)_3P]_2Pt(CF_2CFX) + 2 (C_6H_5)_3P \quad [43] \quad \text{(b)}$$
$$(X = F, Cl, CF_3) \qquad \text{(XXV)}$$

$$[(C_6H_5)_3P]_4Pt + (CF_3)_2CO \longrightarrow [(C_6H_5)_3P]_2Pt[OC(CF_3)_2] + 2\ (C_6H_5)_3P \quad [43] \qquad (c)$$
$$\textbf{(XXVI)}$$

$$[(C_6H_5)_3P]_4Pt + YCS \longrightarrow [(C_6H_5)_3P]_2Pt(SCY) + 2\ (C_6H_5)_3P \quad [44] \qquad (d)$$
$$\textbf{(XXVII)}$$

These reactions proceed easily and in good yield at room temperature.

XXIV

XXV

XXVI

XXVII

REFERENCES

1. W. R. Gilliland and A. A. Blanchard, *Inorg. Syn.* **2**, 234 (1946).
2. L. Mond, C. Langer, and F. Quincke, *J. Chem. Soc.* **57**, 749 (1890).
3. W. Hieber, E. O. Fischer, and E. Böckly, *Z. Anorg. Allgem. Chem.* **269**, 308 (1952).
4. E. O. Fischer and R. Jira, *Z. Naturforsch.* **8b**, 217 (1953).
5. J. F. Cordes, *Ber.* **95**, 3084 (1962).
6. R. B. King, *Organometal. Syn.* **1**, 71 (1965).
7. M. Dubeck and A. H. Filbey, *J. Am. Chem. Soc.* **83**, 1257 (1961).
8. T. S. Piper, F. A. Cotton, and G. Wilkinson, *J. Inorg. & Nucl. Chem.* **1**, 165 (1965).
9. W. K. Schropp, *J. Inorg. & Nucl. Chem.* **24**, 1688 (1962).
10. E. O. Fischer and C. Palm, *Ber.* **91**, 1725 (1958).
11. H. Yamazaki, T. Nishido, Y. Matsumoto, S. Sumida, and N. Hagihara, *J. Organometal. Chem. (Amsterdam)* **6**, 86 (1966).
12. J. R. Olechowski, C. G. MacAlister, and R. F. Clark, *Inorg. Chem.* **4**, 246 (1965); H. Behrens and K. Meyer, *Z. Naturforsch.* **21b**, 489 (1966).
13. M. Dubeck, *J. Am. Chem. Soc.* **82**, 502 and 6193 (1960).
14. D. W. McBride, E. Dudek, and F. G. A. Stone, *J. Chem. Soc.* p. 1725 (1964).
15. D. J. Patmore and W. A. G. Graham, *Inorg. Chem.* **5**, 1405 (1966).
16. B. Bogdanović, M. Kröner, and G. Wilke, *Ann.* **699**, 1 (1966).
17. G. N. Schrauzer and H. Thyret, *J. Am. Chem. Soc.* **82**, 6420 (1960).
18. G. N. Schrauzer and H. Thyret, *Z. Naturforsch.* **16b**, 353 (1961).
19. R. Criegee and G. Schröder, *Ann.* **623**, 1 (1959).
20. W. E. Oberhansli and L. F. Dahl, *Inorg. Chem.* **4**, 150 (1965).
21. R. B. King, M. I. Bruce, J. R. Phillips, and F. G. A. Stone, *Inorg. Chem.* **5**, 684 (1966).
22. M. J. Bennett, F. A. Cotton, and B. H. C. Winquist, *J. Am. Chem. Soc.* **89**, 5366 (1967).
23. G. N. Schrauzer, *J. Am. Chem. Soc.* **81**, 5310 (1959).

24. J. K. Becconsall, B. E. Job, and S. O'Brien, *J. Chem. Soc., A* p. 423 (1967).
25. H. Bönnemann, B. Bogdanović, and G. Wilke, *Angew. Chem. Intern. Ed. Engl.* **6**, 804 (1967).
26. E. O. Fischer and G. Bürger, *Ber.* **94**, 2409 (1961).
27. J. Smidt and W. Hafner, *Angew. Chem.* **71**, 284 (1959).
28. W. R. McClellan, H. H. Hoehn, H. N. Cripps, E. L. Muetterties, and B. W. Howk, *J. Am. Chem. Soc.* **83**, 1601 (1961).
29. B. L. Shaw, *Proc. Chem. Soc.* p. 247 (1960).
30. B. L. Shaw and N. Sheppard, *Chem. & Ind. (London)* p. 517 (1961).
31. M. E. Foss and C. S. Gibson, *J. Chem. Soc.* p. 299 (1951).
32. S. D. Robinson and B. L. Shaw, *J. Chem. Soc.* p. 1529 (1965).
33. J. Chatt and B. L. Shaw, *J. Chem. Soc.* pp. 705 and 4020 (1959); p. 1718 (1960); G. Calvin and G. E. Coates, *J. Chem. Soc.* p. 2009 (1960).
34. W. C. Zeise, *Pogg. Annalen* [2] **9**, 632 (1827).
35. J. Chatt and L. A. Duncanson, *J. Chem. Soc.* p. 2939 (1953).
36. J. Chatt and R. G. Wilkins, *J. Chem. Soc.* p. 2622 (1952).
37. M. S. Kharasch, R. C. Seyler, and F. R. Mayo, *J. Am. Chem. Soc.* **60**, 882 (1938).
38. J. Smidt, W. Hafner, R. Jira, J. Sedlmeier, R. Sieber, R. Rüttinger, and H. Kojer, *Angew. Chem.* **71**, 176 (1959).
39. J. Chatt, L. M. Vallarino, and L. M. Venanzi, *J. Chem. Soc.* p. 2496 (1957).
40. J. Chatt, L. M. Vallarino, and L. M. Venanzi, *J. Chem. Soc.* p. 3413 (1957).
41. J. Müller and P. Göser, *Angew. Chem. Intern. Ed. Engl.* **6**, 364 (1967).
42. D. A. Harbourne, D. T. Rosevear, and F. G. A. Stone, *Inorg. Nucl. Chem. Letters* **2**, 247 (1966).
43. M. Green, R. B. L. Osborn, A. J. Rest, and F. G. A. Stone, *Chem. Commun.* p. 502 (1966).
44. M. C. Baird and G. Wilkinson, *Chem. Commun.* p. 514 (1966).

SUPPLEMENTARY READING

1. G. N. Schrauzer, Some advances in the organometallic chemistry of nickel. *Advan. Organometal. Chem.* **2**, 3 (1964).
2. G. P. Chiusoli and L. Cassar, Nickel-catalyzed reactions of allyl halides and related compounds. *Angew. Chem. Intern. Ed. Engl.* **6**, 124 (1967).
3. J. R. Miller, Recent advances in the stereochemistry of nickel, palladium, and platinum. *Advan. Inorg. Chem. Radiochem.* **4**, 133 (1962).
4. J. Smidt, W. Hafner, R. Jira, R. Sieber, J. Sedlmeier, and A. Sabel, The oxidation of olefins with palladium chloride catalysts. *Angew. Chem. Intern. Ed. Engl.* **1**, 80 (1962).
5. K. Bittler, N. von Kutepow, D. Neubauer, and H. Reis, Carbonylation of olefins under mild temperature conditions in the presence of palladium complexes. *Angew. Chem. Intern. Ed. Engl.* **7**, 329 (1968).
6. A. Aguiló, Olefin oxidation with palladium(II) catalysts in solution. *Advan. Organometal. Chem.* **5**, 321 (1967).

CHAPTER VIII

Organometallic Derivatives of Copper, Silver, and Gold

Most organometallic compounds of copper, silver, and gold are relatively unstable. Nevertheless, olefin complexes and acetylides are known for all three metals. In addition, alkyl derivatives of copper(I) and silver(I) are known but are unstable, except for some fluorinated organocopper derivatives. Alkyl derivatives of gold(III) are known and appear to be more stable than the alkyl derivatives of copper(I) and silver(I). The only known carbonyl derivatives of these three metals are some rather unstable, poorly understood carbonyl halides of copper and gold.

The alkyl derivatives of copper(I) are obtained by reactions of copper(I) halides with alkyllithium or alkylmagnesium halides in diethyl ether solution [1]. They are unstable solids which readily decompose at around room temperature to give metallic copper. Pentafluorophenylcopper(I), $[C_6F_5Cu]_3$, which may be prepared from pentafluorophenylmagnesium bromide and copper(I) halides, is considerably more stable, since it can be heated to 130° in vacuo without decomposition [2]. Another relatively stable organocopper compound is the colorless cyclopentadienyl derivative $C_5H_5CuP(C_2H_5)_3$, mp 127°–128°, which may be obtained from a mixture of copper(I) oxide, triethylphosphine, and cyclopentadiene in an inert solvent [3]. This cyclopentadienyl derivative is a fluxional molecule, since its proton NMR spectrum is temperature dependent [4].

The alkyl derivatives of silver(I) are even less stable than those of copper(I) [5]. They are formed by treatment of silver nitrate with tetraalkyllead derivatives at 0°. Upon warming above this temperature they decompose readily into silver metal and the free organic radical; the latter then decomposes, giving various hydrocarbons. Thus, decomposition of methylsilver proceeds cleanly, to give metallic silver and ethane.

The chemistry of alkyl and aryl derivatives of gold(III) is somewhat more extensive. Treatment of gold(III) bromide with methyllithium at −65° gives trimethylgold, $(CH_3)_3Au$, which is so unstable that it decomposes above −40°. However, complexing trimethyl gold with an amine (e.g., benzylamine or α-aminopyridine) gives adducts of the type $(CH_3)_3AuL$ (L = amine), which form crystalline solids stable at room temperature [6]. Thus, increasing the apparent coordination number of gold(III) from 3 to 4 appears to greatly stabilize its organometallic derivatives. Reaction of methylmagnesium iodide with various gold(III) halide derivatives {a good one is the pyridine complex $[(C_5H_5N)_2AuCl_2]_2Cl$} gives colorless $[(CH_3)_2AuI]_2$, mp 78.5°, of structure I with iodine bridges and 4-coordinate Au(III) [7]. This methylgold derivative explodes on shock, especially above its melting point. By reactions with various amines or nucleophiles $[(CH_3)_2AuI]_2$ can be converted into other dimethyl-gold(III) derivatives. All of these compounds contain planar 4-coordinate gold(III). Thus, ethylenediamine reacts with $[(CH_3)_2AuI]_2$ to give both the salt $[(NH_2CH_2CH_2NH_2)Au(CH_3)_2]I$ of structure II (R = CH_3), where the

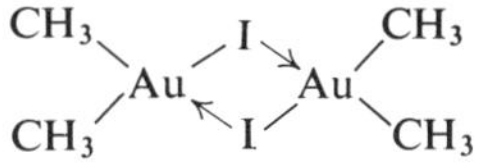

I

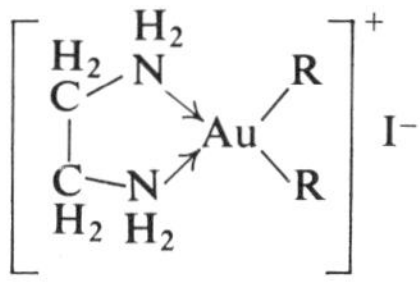

II

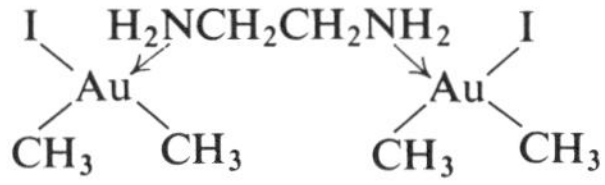

III

ethylenediamine acts as a bidentate chelating ligand, and the nonionic deri-vative $[(CH_3)_2AuI]_2(NH_2CH_2CH_2NH_2)$ of structure III (R = CH_3) where the ethylenediamine acts as a bridging ligand. Similarly, thallium (I) acetyl-acetonate reacts with $[(CH_3)_2AuI]_2$ to give the mononuclear acetylacetonate $(CH_3)_2AuC_5H_7O_2$ (IV). Metal cyanides react with $[(CH_3)_2AuI]_2$ to give the tetranuclear cyanide derivative $[(CH_3)_2AuCN]_4$ (V). The tetranuclear formu-lation of V arises naturally from the square planar geometry of the 4-coordinate Au(III) and the linear sp hybridization of both the carbon and nitrogen atoms in the CN groups.

Compounds analogous to these dimethylgold(III) derivatives containing other alkyl groups such as ethyl, and propyl, have been prepared. Analogous aryl derivatives of gold(III) have not been prepared. However, arylgold(III)

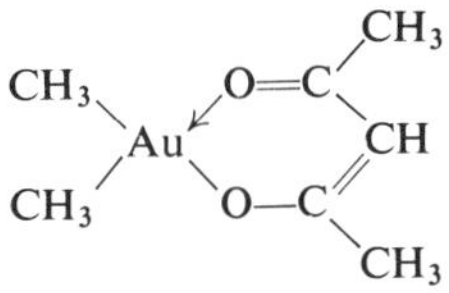

IV

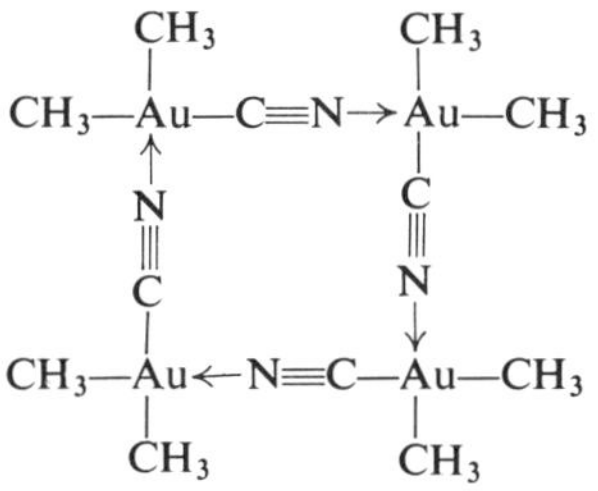

V

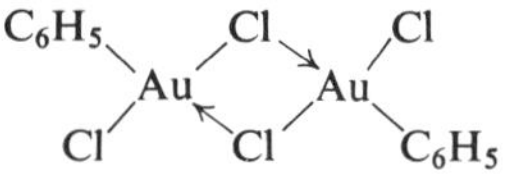

VI

derivatives of other types can be prepared by the so-called *auration* (analogous to mercuration) of aromatic compounds. Thus, treatment of benzene with gold(III) chloride in the presence of a weak base such as diethyl ether gives pale yellow, crystalline $[C_6H_5AuCl_2]_2$ (**VI**) [*8*]. The weak base is necessary for this auration reaction, since in its absence benzene and gold(III) chloride give gold(I) chloride and chlorinated benzenes, rather than organogold derivatives.

REFERENCES

1. H. Gilman and J. M. Straley, *Rec. Trav. Chim.* **55**, 821 (1936); H. Gilman, R. G. Jones, and L. A. Woods, *J. Org. Chem.* **17**, 1630 (1952).
2. A. Cairncross and W. A. Sheppard, *J. Am. Chem. Soc.* **90**, 2186 (1968).
3. G. Wilkinson and T. S. Piper, *J. Inorg. & Nucl. Chem.* **2**, 32 (1956).
4. G. M. Whitesides and J. S. Fleming, *J. Am. Chem. Soc.* **89**, 2855 (1967).
5. G. Costa and A. Camus, *Gazz. Chim. Ital.* **86**, 77 (1956).
6. H. Gilman and L. A. Woods, *J. Am. Chem. Soc.* **70**, 550 (1948).
7. F. H. Brain and C. S. Gibson, *J. Chem. Soc.* p. 762 (1939).
8. M. S. Kharasch and H. S. Isbell, *J. Am. Chem. Soc.* **53**, 3053 (1931).

Supplementary Reading

Physical Studies on Transition-Metal Organometallic Compounds

1. M. R. Churchill and R. Mason, The structural chemistry of organo-transition metal complexes: Some recent developments. *Advan. Organometal. Chem.* **5**, 93 (1967).
2. H. P. Fritz, Infrared and Raman spectral studies of π-complexes formed between metals and C_nH_n rings. *Advan. Organometal. Chem.* **1**, 256 (1964).
3. M. I. Bruce, Mass spectra of organometallic compounds. *Advan. Organometal. Chem.* **6**, 273 (1968).
4. J. Lewis and B. F. G. Johnson, Mass spectra of some organometallic molecules. *Accounts Chem. Res.* **1**, 245 (1968).
5. M. L. Maddox, S. L. Stafford, and H. D. Kaesz, Applications of nuclear magnetic resonance to the study of organometallic compounds. *Advan. Organometal. Chem.* **3**, 4 (1965).
6. F. A. Cotton, Fluxional organometallic molecules. *Accounts Chem. Res.* **1**, 257 (1968).

Transition-Metal Organometallic Compounds in Organic Synthesis

1. C. W. Bird, "Transition Metal Intermediates in Organic Synthesis." Logos Press, London, 1967.
2. C. W. Bird, Synthesis of organic compounds by direct carbonylation reactions using metal carbonyls. *Chem. Rev.* **62**, 283 (1962).
3. G. Wilke, Cyclooligomerization of butadiene and transition-metal π-complexes. *Angew. Chem. Intern. Ed. Engl.* **2**, 105 (1963).
4. J. P. Collman, Patterns of organometallic reactions related to homogeneous catalysis. *Accounts Chem. Res.* **1**, 136 (1968).

Author Index

Numbers in parentheses are reference numbers and indicate that an author's work is referred to although his name is not cited in the text. Numbers in italic show the page on which the complete reference is listed.

A

Abeck, W., 150(9), *163*
Abel, E. W., *42*, 65(7), 79(45), *89, 90,* 159(70), *164*
Aguiló, A., *177*
Åkermark, B., 50(24a), *61*
Alberola, A., 54(39), *61*
Alexander, L. E., 149(4), *163*
Allegra, G., 159(71), *165*
Amiet, R. G., 21(33), *41*
Andrews, T. D., 38(58), *42*
Anet, F. A. L., 89(83, 84), *91*
Angelici, R. J., 156(55), *164*
Anisimov, K. N., 13(20), 14(22), *40,* 64 (5), *89*
Appelbaum, M., 116(28), *145*
Araneo, A., 124(50), *145,* 152(25), *163*
Arnet, J. E., 125(55), *145*
Augl, J. M., 31(48), *41,* 69(22), 72(33), *90*
Aylett, B. J., 151(17), *163*

B

Bader, G., 120(38), *145*
Bailey, M. F., 94(1), *109*
Baird, M. C., 176(44), *177*
Banks, R. E., 135(79), *146*
Barber, W. A., 19(24), *40*
Barnerjee, A. K., 114(14), *144*
Barnett, K. W., 80(47), *90*
Barraclough, C. G., 98(17), *109*
Bartlett, P. D., 49(20), *61*
Basolo, F., 98(16), *109,* 151(19), *163*
Becconsall, J. K., 53(37), *61,* 161(75), *165,* 172(24), *177*
Beck, W., 13(17), *40,* 119(37), 124 (50), *145,* 149(7), *163*
Beckert, O., 81(53), *90*
Beerman, C., 50(25), *61*
Behrens, H., 66((8), 67(13), 86(71), *89, 91,* 169(12), *176*

Bennett, M. A., *42,* 159(70), *164*
Bennett, M. J., 137(87), 142(105), *146, 147,* 171(22), *176*
Berlin, A. M., 48(15), *61*
Bernardi, G., 13(14), *40*
Bertelli, D. J., 126(57), 127(57), *146*
Berthold, H. J., 50(26), *61*
Bestian, H., 50(25), *61, 62*
Beutner, H., 123(46), 124(46), *145*
Bibler, J. P., 141(102, 103), *147*
Bird, C. W., *181*
Birmingham, J. M., 19(23), *40,* 44(10b), *60,* 84(61), *91,* 100(20), *109,* 155 (49), *164*
Bisnette, M. B., 71(28), 76(39), 78(43), 79(44), 81(52), 83(59), 87(74), *90, 91,* 105(36), 106(36, 39), *110,* 117(30), 122(43), 123(43), 139 (95), 140(99), 141(101), 142 (108), *145, 147,* 151(22), *163*
Bittler, K., 157(58), *164, 177*
Blanchard, A. A., 12(10), *40,* 166(1), 169(1), 170(1), *176*
Boche, G., 50(24a), *61*
Böckly, E., 150(12), *163,* 167(3), *176*
Bogdanovic, B., 31(50), *41, 42,* 170(16), 172(25), *176*
Boleseva, I. N., 143(113), *147*
Bonati, F., 139(97), *147,* 154(39), *164*
Bönnemann, H., 172(25), *177*
Bor, G., 148(1), 152(26), *163*
Boss, C. R., 131(67), *146*
Böttcher, R., 143(109, 110), *147*
Bowden, F. L., 133(74), *146*
Brack, A., 113(7), *144*
Bradford, C. W., 12(11), *40,* 137(88), *146*
Brain, F. H., 179(7), *180*
Brainina, E. M., 51(31), *61*
Braun, G., 119(37), *145,* 150(8, 14), *163*

183

Subject Index

A

Acetaldehyde, 175
Acetonitrile, 69
    π-complexes of, 22, 70
    as ligand, 64, 72
Acetylacetone, 159
Acetyl chloride, 96, 114, 150
Acetyl derivatives, 106
Acetylene(s), 72, 133–134
    as ligand, 9
    trimerization, 24
Acetylenic esters, 101
Acetylferrocene, 114
Acetylides, 178
Acrolein, 125
Acrylamide, 125
Acrylic acid, 125
    derivatives of, 27
Acrylonitrile, 27, 125, 171
Actinides, organometallic chemistry of,
    45–46
Acyl chlorides, 96, 104
Acyl ferrocenes, 114
Acyl groups as ligands, 8
Acyl halides, 114, 140
Alcohols, 150
Aldehydes, 175
Alkali metal borohydrides, 68
Alkylaluminum compounds, 161
Alkylferrocenylmethyl sulfide, 114
Alkyl groups as ligands, 8, 35
Alkyllithium, 48, 174, 178
Alkylmagnesium compounds, 163
Alkylmagnesium halides, 174, 178
    as reducing agents, 12, 13, 64, 104
Alkyl nitrate, 115
Alkylthiochloroformates, 114
Alkylthioferrocenyl carboxylates, 114
Alkyltitanium, 48, 50
Alkyne-dicobalt hexacarbonyls, 152, 153

Alkynes, 72, 152, 169
    as ligands, 57
    metal carbonyl derivatives of, 68–75
Allenes, 131–133
Allyl alcohol, 172
Allyl bromide, 151
Allyl chloride, 77, 124, 160
π-Allylcobalt derivatives, 160
π-Allyl-π-1,5-cyclooctadiene-rhodium, 161
π-Allyl derivatives, 78
    "isoleptic," 31, 33
π-Allyl group as ligand, 8
Allyl halides, 32, 66, 131
π-Allylic cobalt tricarbonyls, 151
π-Allyliron carbonyl derivatives, 131
π-Allyliron tricarbonyl halides, 131
Allylmagnesium bromide, 88
Allylmagnesium chloride, 53, 60, 161,
    172, 173
Allylmagnesium halides, 34
Allylnickel halides, 172
π-Allylrhodium derivatives, 160
π-Allyltricarbonylcobalt, 151
Aluminum, 108
    powdered, 23, 24, 58, 60, 85, 143
    as reducing agent, 12, 13
Aluminum alkyls, 50
    as reducing agents, 12, 13
Aluminum chloride, 23, 58, 60, 80, 83,
    85, 104, 107, 108, 113, 114, 143
Aluminum halides as catalysts, 23, 80,
    143
Amines, 83, 150
    as ligands, 73, 105
Aminoferrocene, 115
α-Aminopyridine, 178
Ammonia
    liquid, 19
    reaction with lanthanides, 44
Ammonium hexafluorophosphate, 23
Ammonium ion in nitrogen fixation, 50
n-Amylsodium, 86